PARTIAL SOLUTIONS MANUAL

to accompany

ELEMENTARY ALGEBRA
For College Students,
Third Edition

Jerome E. Kaufmann

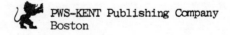

PWS-KENT Publishing Company
Boston

PWS-KENT
Publishing Company

20 Park Plaza
Boston, Massachusetts 02116

PWS-KENT Publishing Company is a division of Wadsworth, Inc.

Printed in the United States of America.

89 90 91 92 93 - 10 9 8 7 6 5 4 3 2 1

ISBN 0-534-91632-5

PREFACE

This student supplement has been prepared to accompany the text Third Edition of Elementary Algebra for College Students. Detailed solutions are included for problems numbered 1,5,9,13,... etc. .

Many of the exercises can be solved in more than one way. However, in order to keep this manual a convenient size, I have usually shown only one possible approach. You may devise other approaches that are as simple as the ones in this manual.

This manual has been constructed for your convenience. Please use it in the way that it best suits your needs. However, I would suggest that you refer to it only after you have made a sincere effort to solve the problems by yourself.

Good luck in your study of elementary algebra.

J. Kaufmann

CONTENTS

Chapter 1..1

Chapter 2...5

Chapter 3..10

Chapter 4..22

Chapter 5..34

Chapter 6..41

Chapter 7..53

Chapter 8..63

Chapter 9..74

Chapter 10..79

Chapter 11..89

Problem Set 1.1

1. $9+14-7 = 23-7 = 16$

5. $16+5 \cdot 7 = 16+35 = 51$

9. $4(7)+6(9) = 28+54 = 82$

13. $(6+9)(8-4) = (15)(4) = 60$

17. $16 \div 8 \cdot 4 + 36 \div 4 \cdot 2 = 2 \cdot 4 + 9 \cdot 2 = 8+18 = 26$

21. $56-[3(9-6)] = 56-[3(3)] = 56-9 = 47$

25. $32 \div 8 \cdot 2 + 24 \div 6 - 1 = 4 \cdot 2 + 4 - 1 = 8 + 4 - 1 = 11$

29. $\dfrac{6(8-3)}{3} + \dfrac{12(7-4)}{9} = \dfrac{6(5)}{3} + \dfrac{12(3)}{9} = \dfrac{30}{3} + \dfrac{36}{9} = 10+4 = 14$

33. $\dfrac{4 \cdot 6 + 5 \cdot 3}{7 + 2 \cdot 3} + \dfrac{7 \cdot 9 + 6 \cdot 5}{3 \cdot 5 + 8 \cdot 2} = \dfrac{24+15}{7+6} + \dfrac{63+30}{15+16} = \dfrac{39}{13} + \dfrac{93}{31} = 3+3 = 6$

37. $16a-9b = 16(3)-9(4) = 48-36 = 12$

41. $14xz+6xy-4yz = 14(8)(7)+6(8)(5)-4(5)(7) = 784+240-140 = 884$

45. $\dfrac{y+16}{6} + \dfrac{50-y}{3} = \dfrac{8+16}{6} + \dfrac{50-8}{3} = \dfrac{24}{6} + \dfrac{42}{3} = 4+14 = 18$

49. $(5x-2y)(3x+4y) = [5(3)-2(6)][3(3)+4(6)]$
$$= [15-12][9+24] = (3)(33) = 99$$

53. $81-2[5(n+4)] = 81-2[5(3+4)]$
$\quad = 81-2[5(7)]$
$\quad = 81-2(35)$
$\quad = 81-70$
$\quad = 11$

57. $\dfrac{bh}{2} = \dfrac{7(6)}{2} = \dfrac{42}{2} = 21$

61. $\dfrac{Bh}{3} = \dfrac{27(9)}{3} = 81$

65. $\dfrac{Bh}{3} = \dfrac{36(7)}{3} = 84$

69. $\dfrac{h(b_1+b_2)}{2} = \dfrac{8(17+24)}{2} = 4(41) = 164$

Problem Set 1.2

1. Since $8(7) = 56$, "8 divides 56" is a true statement.

5. Since $8(12) = 96$, "96 is a multiple of 8" is a true statement.

9. Since $4(36) = 144$, "144 is divisible by 4" is a true statement.

13. Since $11(13) = 143$, "11 is a factor of 143" is a true statement.

17. Since $(3)(19) = 57$ and 3 is a prime number, "3 is a prime factor of 57" is a true statement.

21. Since 53 is only divisible by itself and 1, it is a prime number.

25. Since $91 = 7(13)$, it is a composite number.

29. Since $111 = 3(37)$, it is a composite number.

33. $36 = 4 \cdot 9 = 2 \cdot 2 \cdot 3 \cdot 3$

37. $56 = 8 \cdot 7 = 2 \cdot 2 \cdot 2 \cdot 7$

41. $135 = 5 \cdot 27 = 3 \cdot 3 \cdot 3 \cdot 5$

45. $56 = 2 \cdot 2 \cdot 2 \cdot 7$
     $64 = 2 \cdot 2 \cdot 2 \cdot 2 \cdot 2 \cdot 2$ $\longrightarrow$ The greatest common factor is $2 \cdot 2 \cdot 2 = 8$.

49. $84 = 2 \cdot 2 \cdot 3 \cdot 7$
     $96 = 2 \cdot 2 \cdot 2 \cdot 2 \cdot 2 \cdot 3$ $\longrightarrow$ The greatest common factor is $2 \cdot 2 \cdot 3 = 12$.

53. $48 = 2 \cdot 2 \cdot 2 \cdot 2 \cdot 3$
     $60 = 2 \cdot 2 \cdot 3 \cdot 5$ $\longrightarrow$ The greatest common factor is $2 \cdot 2 \cdot 3 = 12$.
     $84 = 2 \cdot 2 \cdot 3 \cdot 7$

57. $12 = 2 \cdot 2 \cdot 3$
     $16 = 2 \cdot 2 \cdot 2 \cdot 2$ $\longrightarrow$ The least common multiple is $2 \cdot 2 \cdot 2 \cdot 2 \cdot 3 = 48$.

61. $49 = 7 \cdot 7$
     $56 = 2 \cdot 2 \cdot 2 \cdot 7$ $\longrightarrow$ The least common multiple is $2 \cdot 2 \cdot 2 \cdot 7 \cdot 7 = 392$.

65. $9 = 3 \cdot 3$
     $15 = 3 \cdot 5$ $\longrightarrow$ The least common multiple is $2 \cdot 3 \cdot 3 \cdot 5 = 90$.
     $18 = 2 \cdot 3 \cdot 3$

Problem Set 1.3

1.

$5+(-3) = 2$

5.

$-3+(-4) = -7$

9.

$5+(-11) = -6$

13. $8+(-19) = -(|-19|-|8|) = -(19-8) = -11$

17. $-15+8 = -(|-15|-|8|) = -(15-8) = -7$

21. $-27+8 = -(|-27|-|8|) = -(27-8) = -19$

25. $-25+(-36) = -(|-25|+|-36|) = -(25+36) = -61$

29. $-34+(-58) = -(|-34|+|-58|) = -(34+58) = -92$

33. $-4-9 = -4+(-9) = -13$ 37. $-6-(-12) = -6+12 = 6$

41. $-18-27 = -18+(-27) = -45$ 45. $45-18 = 45+(-18) = 27$

2

49. $-53-(-24) = -53+24 = -29$ 53. $-4-(-6)+5-8 = -4+6+5+(-8) = -1$

57. $-6-4-(-2)+(-5) = -6+(-4)+2+(-5) = -13$

61. $7-12+14-15-9 = 7+(-12)+14+(-15)+(-9) = -15$

65. $16-21+(-15)-(-22) = 16+(-21)+(-15)+22 = 2$

69. $\begin{array}{r} -13 \\ \underline{-18} \\ -31 \end{array}$ This problem in horizontal format is $(-13)+(-18) = -31$.

73. $\begin{array}{r} -21 \\ \underline{39} \\ 18 \end{array}$ This problem in horizontal format is $-21+39 = 18$.

77. $\begin{array}{r} -53 \\ \underline{24} \\ -29 \end{array}$ This problem in horizontal format is $-53+24 = -29$.

81. $\begin{array}{r} 6 \\ (+)\underline{-9} \\ 15 \end{array}$ Change the sign of the bottom number and add.

85. $\begin{array}{r} 17 \\ (+)\underline{-19} \\ 36 \end{array}$ Change the sign of the bottom number and add.

89. $\begin{array}{r} -12 \\ (-)\underline{12} \\ -24 \end{array}$ Change the sign of the bottom number and add.

93. $-x+y-z = -3+(-4)-(-6) = -3+(-4)+6 = -1$

97. $-x+y+z = -(-11)+7+(-9) = 11+7+(-9) = 9$

Problem Set 1.4

1. $5(-6) = -(|5|\cdot|-6|)-(5\cdot6) = -30$

5. $\dfrac{-42}{-6} = -\dfrac{|-42|}{|-6|} = \dfrac{42}{6} = 7$ 9. $(-5)(-12) = |-5|\cdot|-12| = 5\cdot12 = 60$

13. $14(-9) = -(|14|\cdot|-9|) = -(14\cdot9) = -126$

17. $\dfrac{135}{-15} = -(\dfrac{|135|}{|-15|}) = -(\dfrac{135}{15}) = -9$ 21. $(-15)(-15) = |-15|\cdot|-15| = 15\cdot15 = 225$

25. $\dfrac{0}{-8}$ because $(-8)(0) = 0$. 29. $\dfrac{76}{-4} = -(\dfrac{|76|}{|-4|}) = -(\dfrac{76}{4}) = -19$

33. $(-56)\div(-4) = |-56|\div|-4| = 56\div4 = 14$

37. $(-72)\div18 = -(|-72|\div|18|) = -(72\div18) = -4$

41. $3(-4)+5(-7) = -12+(-35) = -47$ 45. $(-3)(-8)+(-9)(-5) = 24+45 = 69$

49. $\dfrac{13+(-25)}{-3} = \dfrac{-12}{-3} = 4$ 53. $\dfrac{-7(10)+6(-9)}{-4} = \dfrac{-70+(-54)}{-4} = \dfrac{-124}{-4} = 31$

57. $-2(3)-3(-4)+4(-5)-6(-7) = -6+12+(-20)+42 = 28$

61. $7x+5y = 7(-5)+5(9) = -35+45 = 10$ 65. $-6x-7y = -6(-4)-7(-6) = 24+42 = 66$

69. $3(2a-5b) = 3[2(-1)-5(-5)] = 3[-2+25] = 3(23) = 69$

73. $-4ab-b = -4(2)(-14)-(-14) = 112+14 = 126$

77. $\dfrac{5(F-32)}{9} = \dfrac{5(59-32)}{9} = \dfrac{5(27)}{9} = 15$ 81. $\dfrac{5(F-32)}{9} = \dfrac{5(-13-32)}{9} = \dfrac{5(-45)}{9} = -25$

3

85. $\frac{9C}{5} + 32 = \frac{9(40)}{5} + 32 = 72+32 = 104$

Problem Set 1.5

13. $(-18+56)+18 = (56+(-18))+18$
$= 56+(-18+18) = 56+0 = 56$

17. $(25)(-18)(-4) = (25)(-4)(-18) = (-100)(-18) = 1800$

21. $37(-42-58) = 37(-100) = -3700$ 25. $15(-14)+16(-8) = -210-128 = -338$

29. $-21+22-23+27+21-19 = 7$ 33. $4m+m-8m = (4+1-8)m = -3m$

37. $4x-3y-7x+y = 4x-7x-3y+y$
$= (4-7)x+(-3+1)y$
$= -3x-2y$

41. $6xy-x-13xy+4x = 6xy-13xy-x+4x$
$= -7xy+3x$

45. $-2xy+12+8xy-16 = -2xy+8xy+12-16$
$= 6xy-4$

49. $13ab+2a-7a-9ab+ab-6a =$
$13ab-9ab+ab+2a-7a-6a = 5ab-11a$

53. $5(x-4)+6(x+8) = 5x-20+6x+48$
$= 11x+28$

57. $3(a-1)-2(a-6)+4(a+5) = 3a-3-2a+12+4a+20$
$= 5a+29$

61. $(y+3)-(y-2)-(y+6)-7(y-1) = y+3-y+2-y-6-7y+7$
$= -8y+6$

65. $5(x-2)+8(x+6) = 5x-10+8x+48$
$= 13x+38$
$= 13(-6)+38$
$= -78+38 = -40$

69. $(x-6)-(x+12) = x-6-x-12 = -18$

73. $2xy+6+7xy-8 = 9xy-2$
$= 9(2)(-4)-2$
$= -72-2 = -74$

77. $(a-b)-(a+b) = a-b-a-b = -2b$
$= -2(-17)$
$= 34$

4

Chapter 2

Problem Set 2.1

1. $\dfrac{8}{12} = \dfrac{4 \cdot 2}{4 \cdot 3} = \dfrac{2}{3}$

5. $\dfrac{15}{9} = \dfrac{3 \cdot 5}{3 \cdot 3} = \dfrac{5}{3}$

9. $\dfrac{27}{-36} = -\dfrac{9 \cdot 3}{9 \cdot 4} = -\dfrac{3}{4}$

13. $\dfrac{24x}{44x} = \dfrac{4x \cdot 6}{4x \cdot 11} = \dfrac{6}{11}$

17. $\dfrac{14xy}{35y} = \dfrac{7y \cdot 2x}{7y \cdot 5} = \dfrac{2x}{5}$

21. $\dfrac{-56yz}{-49xy} = \dfrac{-7y \cdot 8z}{-7y \cdot 7x} = \dfrac{8z}{7x}$

25. $\dfrac{3}{4} \cdot \dfrac{5}{7} = \dfrac{3 \cdot 5}{4 \cdot 7} = \dfrac{15}{28}$

29. $\dfrac{3}{8} \cdot \dfrac{12}{15} = \dfrac{\overset{1}{\cancel{3}} \cdot \overset{3}{\cancel{12}}}{\underset{2}{\cancel{8}} \cdot \underset{5}{\cancel{15}}} = \dfrac{3}{10}$

33. $\dfrac{7}{9} \div \dfrac{5}{9} = \dfrac{7}{9} \cdot \dfrac{9}{5} = \dfrac{9 \cdot 7}{9 \cdot 5} = \dfrac{7}{5}$

37. $\left(-\dfrac{8}{10}\right)\left(-\dfrac{10}{32}\right) = \dfrac{\overset{1}{\cancel{8}} \cdot \overset{1}{\cancel{10}}}{\underset{1}{\cancel{10}} \cdot \underset{4}{\cancel{32}}} = \dfrac{1}{4}$

41. $\dfrac{5x}{9y} \cdot \dfrac{7y}{3x} = \dfrac{5 \cdot 7 \cdot \cancel{x} \cdot \cancel{y}}{9 \cdot 3 \cdot \cancel{x} \cdot \cancel{y}} = \dfrac{35}{27}$

45. $\dfrac{10x}{-9y} \cdot \dfrac{15}{20x} = -\dfrac{\overset{1}{\cancel{10}} \cdot \overset{5}{\cancel{15}} \cdot \cancel{x}}{\underset{3}{\cancel{9}} \cdot \underset{2}{\cancel{20}} \cdot \cancel{x} \cdot y} = -\dfrac{5}{6y}$

49. $\left(-\dfrac{7x}{12y}\right)\left(-\dfrac{24y}{35x}\right) = \dfrac{\overset{1}{\cancel{7}} \cdot \overset{2}{\cancel{24}} \cdot \cancel{x} \cdot \cancel{y}}{\underset{1}{\cancel{12}} \cdot \underset{5}{\cancel{35}} \cdot \cancel{x} \cdot \cancel{y}} = \dfrac{2}{5}$

53. $\dfrac{5x}{9y} \div \dfrac{13x}{36y} = \dfrac{5x}{9y} \cdot \dfrac{36y}{13x} = \dfrac{5 \cdot \overset{4}{\cancel{36}} \cdot \cancel{x} \cdot \cancel{y}}{\underset{1}{\cancel{9}} \cdot 13 \cdot \cancel{x} \cdot \cancel{y}} = \dfrac{20}{13}$

57. $\dfrac{-4}{n} \div \dfrac{-18}{n} = \left(-\dfrac{\overset{2}{\cancel{4}}}{\underset{1}{\cancel{n}}}\right)\left(-\dfrac{\overset{1}{\cancel{n}}}{\underset{9}{\cancel{18}}}\right) = \dfrac{2}{9}$

61. $\left(-\dfrac{3}{8}\right)\left(\dfrac{13}{14}\right)\left(-\dfrac{12}{9}\right) = \dfrac{\overset{1}{\cancel{3}} \cdot 13 \cdot \overset{\overset{1}{\cancel{4}}}{\cancel{12}}}{\underset{2}{\cancel{8}} \cdot 14 \cdot \underset{\underset{1}{\cancel{3}}}{\cancel{9}}} = \dfrac{13}{28}$

65. $\left(-\dfrac{2}{3}\right)\left(\dfrac{3}{4}\right) \div \dfrac{1}{8} = \left(-\dfrac{1}{2}\right) \div \dfrac{1}{8} = -\dfrac{1}{2} \cdot 8 = -4$

69. $\left(-\dfrac{6}{7}\right) \div \left(\dfrac{5}{7}\right)\left(-\dfrac{5}{6}\right) = \left(-\dfrac{6}{7}\right)\left(\dfrac{7}{5}\right)\left(-\dfrac{5}{6}\right) = \dfrac{6 \cdot 7 \cdot 5}{7 \cdot 5 \cdot 6} = 1$

73. $\left(\dfrac{5}{2}\right)\left(\dfrac{2}{3}\right) \div \left(-\dfrac{1}{4}\right) \div (-3) = \dfrac{5}{3} \div \left(-\dfrac{1}{4}\right) \div (-3) = \dfrac{5}{3} \cdot (-4) \cdot \left(-\dfrac{1}{3}\right) = \dfrac{20}{9}$

Problem Set 2.2

1. $\dfrac{2}{7} + \dfrac{3}{7} = \dfrac{2+3}{7} = \dfrac{5}{7}$

5. $\dfrac{3}{4} + \dfrac{9}{4} = \dfrac{3+9}{4} = \dfrac{12}{4} = 3$

9. $\dfrac{1}{8} - \dfrac{5}{8} = \dfrac{1-5}{8} = \dfrac{-4}{8} = -\dfrac{1}{2}$

13. $\dfrac{8}{x} + \dfrac{7}{x} = \dfrac{8+7}{x} = \dfrac{15}{x}$

17. $\dfrac{1}{3} + \dfrac{1}{5} = \left(\dfrac{5}{5}\right)\left(\dfrac{1}{3}\right) + \left(\dfrac{3}{3}\right)\left(\dfrac{1}{5}\right) = \dfrac{5}{15} + \dfrac{3}{15} = \dfrac{8}{15}$

21. $\dfrac{7}{10} + \dfrac{8}{15} = \left(\dfrac{3}{3}\right)\left(\dfrac{7}{10}\right) + \left(\dfrac{2}{2}\right)\left(\dfrac{8}{15}\right) = \dfrac{21}{30} + \dfrac{16}{30} = \dfrac{37}{30}$

25. $\dfrac{5}{18} - \dfrac{13}{24} = \left(\dfrac{4}{4}\right)\left(\dfrac{5}{18}\right) - \left(\dfrac{3}{3}\right)\left(\dfrac{13}{24}\right) = \dfrac{20}{72} - \dfrac{39}{72} = -\dfrac{19}{72}$

29. $-\frac{2}{13} - \frac{7}{39} = \left(\frac{3}{3}\right)\left(-\frac{2}{13}\right) - \frac{7}{39} = -\frac{6}{39} - \frac{7}{39} = -\frac{13}{39} = -\frac{1}{3}$

33. $\frac{3}{x} + \frac{4}{y} = \left(\frac{y}{y}\right)\left(\frac{3}{x}\right) + \left(\frac{x}{x}\right)\left(\frac{4}{y}\right) = \frac{3y}{xy} + \frac{4x}{xy} = \frac{3y+4x}{xy}$

37. $\frac{2}{x} + \frac{7}{2x} = \left(\frac{2}{2}\right)\left(\frac{2}{x}\right) + \frac{7}{2x} = \frac{4}{2x} + \frac{7}{2x} = \frac{11}{2x}$

41. $\frac{1}{x} - \frac{7}{5x} = \left(\frac{5}{5}\right)\left(\frac{1}{x}\right) - \frac{7}{5x} = \frac{5}{5x} - \frac{7}{5x} = -\frac{2}{5x}$

45. $\frac{5}{12y} - \frac{3}{8y} = \left(\frac{2}{2}\right)\left(\frac{5}{12y}\right) - \left(\frac{3}{3}\right)\left(\frac{3}{8y}\right) = \frac{10}{24y} - \frac{9}{24y} = \frac{1}{24y}$

49. $\frac{5}{3x} + \frac{7}{3y} = \left(\frac{y}{y}\right)\left(\frac{5}{3x}\right) + \left(\frac{x}{x}\right)\left(\frac{7}{3y}\right) = \frac{5y}{3xy} + \frac{7x}{3xy} = \frac{5y+7x}{3xy}$

53. $\frac{7}{4x} - \frac{5}{9y} = \left(\frac{9y}{9y}\right)\left(\frac{7}{4x}\right) - \left(\frac{4x}{4x}\right)\left(\frac{5}{9y}\right) = \frac{63y}{36xy} - \frac{20x}{36xy} = \frac{63y-20x}{36xy}$

57. $3 + \frac{2}{x} = \left(\frac{x}{x}\right)\left(\frac{3}{1}\right) + \frac{2}{x} = \frac{3x}{x} + \frac{2}{x} = \frac{3x+2}{x}$

61. $\frac{1}{4} - \frac{3}{8} + \frac{5}{12} - \frac{1}{24} = \frac{6}{24} - \frac{9}{24} + \frac{10}{24} - \frac{1}{24} = \frac{6}{24} = \frac{1}{4}$

65. $\frac{3}{4} \cdot \frac{6}{9} - \frac{5}{6} \cdot \frac{8}{10} + \frac{2}{3} \cdot \frac{6}{8} = \frac{1}{2} - \frac{2}{3} + \frac{1}{2}$

$= \frac{3}{6} - \frac{4}{6} + \frac{3}{6}$

$= \frac{2}{6} = \frac{1}{3}$

69. $\frac{4}{5} - \frac{10}{12} - \frac{5}{6} \div \frac{14}{8} + \frac{10}{21} = \frac{4}{5} - \frac{10}{12} - \frac{5}{6} \cdot \frac{8}{14} + \frac{10}{21}$

$= \frac{4}{5} - \frac{10}{12} - \frac{10}{21} + \frac{10}{21}$

$= \frac{4}{5} - \frac{10}{12}$

$= \frac{48}{60} - \frac{50}{60} = -\frac{2}{60} = -\frac{1}{30}$

73. $64\left(\frac{3}{16} + \frac{5}{8} - \frac{1}{4} + \frac{1}{2}\right) = 64\left(\frac{3}{16}\right) + 64\left(\frac{5}{8}\right) - 64\left(\frac{1}{4}\right) + 64\left(\frac{1}{2}\right)$

$= 12 + 40 - 16 + 32 = 68$

77. $\frac{1}{3}x + \frac{2}{5}x = \left(\frac{1}{3} + \frac{2}{5}\right)x = \left(\frac{5}{15} + \frac{6}{15}\right)x = \frac{11}{15}x$

81. $\frac{1}{2}x + \frac{2}{3}x + \frac{1}{6}x = \left(\frac{1}{2} + \frac{2}{3} + \frac{1}{6}\right)x$

$= \left(\frac{3}{6} + \frac{4}{6} + \frac{1}{6}\right)x$

$= \frac{8}{6}x = \frac{4}{3}x$

85. $n + \frac{4}{3}n - \frac{1}{9}n = (1 + \frac{4}{3} - \frac{1}{9})n$

$$= (\frac{9}{9} + \frac{12}{9} - \frac{1}{9})n$$

$$= \frac{20}{9}n$$

89. $\frac{3}{7}x + \frac{1}{4}y + \frac{1}{2}x + \frac{7}{8}y = (\frac{3}{7} + \frac{1}{2})x + (\frac{1}{4} + \frac{7}{8})y$

$$= (\frac{6}{14} + \frac{7}{14})x + (\frac{2}{8} + \frac{7}{8})y$$

$$= \frac{13}{14}x + \frac{9}{8}y$$

Problem Set 2.3

You may find the vertical format helpful for work with decimals.

1. .37
    .25
    ‾‾‾‾
    .62

5. 7.6
    −3.8
    ‾‾‾‾
    3.8

9. 11.3
    − 3.8
    ‾‾‾‾
    7.5

13. $-11.5 - (-10.6) = -11.5 + 10.6$

    −11.5
    10.6
    ‾‾‾‾
    −.9

17. 2.9
    .4
    ‾‾‾‾
    1.16

21. −2.7
    9
    ‾‾‾‾
    −24.3

25. −.13
    −.12
    ‾‾‾‾
    26
    13
    ‾‾‾‾
    .0156

29. $\frac{5.92}{-.8} = \frac{59.2}{-8} = -7.4$

33. 16.5   25.9
    9.4   −18.7
    ‾‾‾‾   ‾‾‾‾
    25.9   7.2

37. $.76(.2 + .8) = .76(1) = .76$

41. $7(.6) + .9 - 3(.4) + .4 = 4.2 + .9 - 1.2 + .4 = 4.3$

45. $5(2.3) - 1.2 - 7.36 \div .8 + .2 = 11.5 - 1.2 - 9.2 + .2 = 1.3$

49. $5.4n - .8n - 1.6n = (5.4 - .8 - 1.6)n = 3n$

53. $3.6x - 7.4y - 9.4x + 10.2y = (3.6 - 9.4)x + (-7.4 + 10.2)y$

$$= -5.8x + 2.8y$$

57. $6(x - 1.1) - 5(x - 2.3) - 4(x + 1.8) = 6x - 6.6 - 5x + 11.5 - 4x - 7.2$

$$= -3x - 2.3$$

61. $x + 2y + 3z = \frac{3}{4} + 2(\frac{1}{3}) + 3(-\frac{1}{6})$

$$= \frac{3}{4} + \frac{2}{3} - \frac{1}{2}$$

$$= \frac{9}{12} + \frac{8}{12} - \frac{6}{12} = \frac{11}{12}$$

65. $-x - 2y + 4z = -1.7 - 2(-2.3) + 4(3.6)$

$$= -1.7 + 4.6 + 14.4$$

$$= 17.3$$

69. $.7x + .6y = .7(-2) + (.6)(6) = -1.4 + 3.6$

$$= 2.2$$

73. $-3a - 1 + 7a - 2 = 4a - 3 = 4(.9) - 3$

$$= 3.6 - 3 = .6$$

7

Problem Set 2.4

1. $2^6 = 2 \cdot 2 \cdot 2 \cdot 2 \cdot 2 \cdot 2 = 64$

5. $(-2)^3 = (-2)(-2)(-2) = -8$

9. $(-4)^2 = (-4)(-4) = 16$

13. $-(\frac{1}{2})^3 = -(\frac{1}{2} \cdot \frac{1}{2} \cdot \frac{1}{2}) = -\frac{1}{8}$

17. $(.3)^3 = (.3)(.3)(.3) = .027$

21. $3^2 + 2^3 - 4^3 = 9 + 8 - 64 = -47$

25. $5(2)^2 - 4(2) - 1 = 20 - 8 - 1 = 11$

29. $-7^2 - 6^2 + 5^2 = -49 - 36 + 25 = -60$

33. $-\frac{3(2)^4}{12} + \frac{5(-3)^3}{15} = \frac{-3(16)}{12} + \frac{5(-27)}{15}$

$$= -4 - 9 = -13$$

37. $3 \cdot 4 \cdot x \cdot y \cdot y = 12xy^2$

41. $(5x)(3y) = 5 \cdot 3 \cdot x \cdot y = 15xy$

45. $(-4a^2)(-2a^3) = (-4)(-2) \cdot a \cdot a \cdot a \cdot a \cdot a = 8a^5$

49. $-12y^3 + 17y^3 - y^3 = (-12 + 17 - 1)y^3 = 4y^3$

53. $\frac{2}{3}n^2 - \frac{1}{4}n^2 - \frac{3}{5}n^2 = (\frac{2}{3} - \frac{1}{4} - \frac{3}{5})n^2$

$$= (\frac{40}{60} - \frac{15}{60} - \frac{36}{60})n^2 = -\frac{11}{60}n^2$$

57. $\frac{22xy^2}{6xy^3} = \frac{\overset{11}{\cancel{22} \cdot \cancel{x} \cdot \cancel{y} \cdot \cancel{y}}}{\underset{3}{\cancel{6} \cdot \cancel{x} \cdot y \cdot \cancel{y} \cdot \cancel{y}}} = \frac{11}{3y}$

61. $\frac{-24abc^2}{32bc} = -\frac{\overset{3}{\cancel{24} \cdot a \cdot \cancel{b} \cdot c \cdot \cancel{c}}}{\underset{4}{\cancel{32} \cdot \cancel{b} \cdot \cancel{c}}} = -\frac{3ac}{4}$

65. $\frac{7x^2}{9y} \cdot \frac{12y}{21x} = \frac{\overset{4}{7 \cdot \cancel{12} \cdot x \cdot \cancel{x} \cdot \cancel{y}}}{\underset{3}{9 \cdot \cancel{21} \cdot \cancel{x} \cdot \cancel{y}}} = \frac{4x}{9}$

69. $\frac{6}{x} + \frac{5}{y^2} = (\frac{y^2}{y^2})(\frac{6}{x}) + (\frac{x}{x})(\frac{5}{y^2})$

$$= \frac{6y^2}{xy^2} + \frac{5x}{xy^2} = \frac{6y^2 + 5x}{xy^2}$$

73. $\frac{3}{2x^3} + \frac{6}{x} = \frac{3}{2x^3} + (\frac{2x^2}{2x^2})(\frac{6}{x})$

$$= \frac{3}{2x^3} + \frac{12x^2}{2x^3} = \frac{3 + 12x^2}{2x^3}$$

77. $\frac{11}{a^2} - \frac{14}{b^2} = (\frac{b^2}{b^2})(\frac{11}{a^2}) - (\frac{a^2}{a^2})(\frac{14}{b^2})$

$$= \frac{11b^2}{a^2b^2} - \frac{14a^2}{a^2b^2} = \frac{11b^2 - 14a^2}{a^2b^2}$$

81. $\frac{3}{x} - \frac{4}{y} - \frac{5}{xy} = (\frac{y}{y})(\frac{3}{x}) - (\frac{x}{x})(\frac{4}{y}) - \frac{5}{xy}$

$$= \frac{3y}{xy} - \frac{4x}{xy} - \frac{5}{xy}$$

$$= \frac{3y - 4x - 5}{xy}$$

85. $3x^2 - y^2 = 3(\frac{1}{2})^2 - (-\frac{1}{3})^2 = 3(\frac{1}{4}) - \frac{1}{9}$

$$= \frac{3}{4} - \frac{1}{9}$$

$$= \frac{27}{36} - \frac{4}{36}$$

$$= \frac{23}{36}$$

89. $-x^2 = -(-8)^2 = -(64) = -64$

93. $-a^2 - 3b^3 = -(-6)^2 - 3(-1)^3$

$$= -36 + 3 = -33$$

For Problems 37-68, it may help to do a specific example before trying to formulate the general expression. Let's illustrate this idea with Problems 37, 49, and 61.

37. Suppose that the sum of two numbers is 35 and one of the numbers is 14. Then to find the other number we subtract 35-14. Thus, if one of the numbers is n, then the other number is 35-n.

49. Suppose that 5 pounds of candy cost $15. Then to find the price per pound we divide 15 by 5. Thus, if p pounds cost d dollars, then d÷p represents the price per pound.

61. To change 15 feet to yards, we divide by 3. Therefore, f feet equals $\frac{f}{3}$ yards.

Chapter 3

Problem Set 3.1

1. $x+9 = 17$
      $x+9-9 = 17-9$ Subtract 9 from both sides.
          $x = 8$

    The solution set is $\{8\}$.

5. $-7 = x+2$
      $-7-2 = x+2-2$ Subtract 2 from both sides.
        $-9 = x$

    The solution set is $\{-9\}$.

9. $21+y = 34$
      $21+y-21 = 34-21$ Subtract 21 from both sides.
              $y = 13$

    The solution set is $\{13\}$.

13. $14 = x-9$
      $14+9 = x-9+9$ Add 9 to both sides.
        $23 = x$

    The solution set is $\{23\}$.

17. $y - \dfrac{2}{3} = \dfrac{3}{4}$

      $y - \dfrac{2}{3} + \dfrac{2}{3} = \dfrac{3}{4} + \dfrac{2}{3}$ Add $\dfrac{2}{3}$ to both sides.

              $y = \dfrac{17}{12}$

    The solution set is $\{\dfrac{17}{12}\}$.

21. $b + .19 = .46$
      $b + .19 - .19 = .46 - .19$ Subtract .19 from both sides.
              $b = .27$

    The solution set is $\{.27\}$.

25. $15-x = 32$
      $15-x-15 = 32-15$ Subtract 15 from both sides.
          $-x = 17$
            $x = -17$ Multiply both sides by -1.

    The solution set is $\{-17\}$.

29. $7x = -56$
      $\dfrac{7x}{7} = \dfrac{-56}{7}$ Divide both sides by 7.
        $x = -8$

    The solution set is $\{-8\}$.

33. $5x = 37$
      $\dfrac{5x}{5} = \dfrac{37}{5}$ Divide both sides by 5.
        $x = \dfrac{37}{5}$

    The solution set is $\{\dfrac{37}{5}\}$.

37. $$-26 = -4n$$

$$\frac{-26}{-4} = \frac{-4n}{-4} \qquad \text{Divide both sides by } -4.$$

$$\frac{13}{2} = n$$

The solution set is $\{\frac{13}{2}\}$.

41. $$\frac{n}{-8} = -3$$

$$-8\left(\frac{n}{-8}\right) = -8(-3) \qquad \text{Multiply both sides by } -8.$$

$$n = 24$$

The solution set is $\{24\}$.

45. $$\frac{3}{4}x = 18$$

$$\frac{4}{3}\left(\frac{3}{4}x\right) = \frac{4}{3}(18) \qquad \text{Multiply both sides by } \frac{4}{3}.$$

$$x = 24$$

The solution set is $\{24\}$.

49. $$\frac{2}{3}n = \frac{1}{5}$$

$$\frac{3}{2}\left(\frac{2}{3}n\right) = \frac{3}{2}\left(\frac{1}{5}\right) \qquad \text{Multiply both sides by } \frac{3}{2}.$$

$$n = \frac{3}{10}$$

The solution set is $\{\frac{3}{10}\}$.

53. $$\frac{3x}{10} = \frac{3}{20}$$

$$\frac{10}{3}\left(\frac{3x}{10}\right) = \frac{10}{3}\left(\frac{3}{20}\right) \qquad \text{Multiply both sides by } \frac{10}{3}.$$

$$x = \frac{1}{2}$$

The solution set is $\{\frac{1}{2}\}$.

57. $$-\frac{4}{3}x = -\frac{9}{8}$$

$$-\frac{3}{4}\left(-\frac{4}{3}x\right) = -\frac{3}{4}\left(-\frac{9}{8}\right) \qquad \text{Multiply both sides by } -\frac{3}{4}.$$

$$x = \frac{27}{32}$$

The solution set is $\{\frac{27}{32}\}$.

61. $$-\frac{5}{7}x = 1$$

$$-\frac{7}{5}\left(-\frac{5}{7}x\right) = -\frac{7}{5}(1) \qquad \text{Multiply both sides by } -\frac{7}{5}.$$

$$x = -\frac{7}{5}$$

The solution set is $\{-\frac{7}{5}\}$.

65. $-8n = \dfrac{6}{5}$

$-\dfrac{1}{8}(-8n) = -\dfrac{1}{8}\left(\dfrac{6}{5}\right)$ Multiply both sides by $-\dfrac{1}{8}$.

$n = -\dfrac{3}{20}$

The solution set is $\{-\dfrac{3}{20}\}$.

Problem Set 3.2

1. $2x+5 = 13$

      $2x+5-5 = 13-5$ Subtract 5 from both sides.

      $2x = 8$

      $\dfrac{2x}{2} = \dfrac{8}{2}$ Divide both sides by 2.

      $x = 4$

The solution set is $\{4\}$.

5. $3x-1 = 23$

      $3x-1+1 = 23+1$ Add 1 to both sides.

      $3x = 24$

      $\dfrac{3x}{3} = \dfrac{24}{3}$ Divide both sides by 3.

      $x = 8$

The solution set is $\{8\}$.

9. $6y-1 = 16$

      $6y-1+1 = 16+1$ Add 1 to both sides.

      $6y = 17$

      $\dfrac{6y}{6} = \dfrac{17}{6}$ Divide both sides by 6.

      $y = \dfrac{17}{6}$

The solution set is $\{\dfrac{17}{6}\}$.

13. $10 = 3t-8$

      $10+8 = 3t-8+8$ Add 8 to both sides.

      $18 = 3t$

      $\dfrac{18}{3} = \dfrac{3t}{3}$ Divide both sides by 3.

      $6 = t$

The solution set is $\{6\}$.

17. $18-n = 23$

      $18-n-18 = 23-18$ Subtract 18 from both sides.

      $-n = 5$

      $-1(-n) = -1(5)$ Multiply both sides by -1.

      $n = -5$

The solution set is $\{-5\}$.

21. $7+4x = 29$

$7+4x-7 = 29-7$ Subtract 7 from both sides.

$4x = 22$

$\dfrac{4x}{4} = \dfrac{22}{4}$ Divide both sides by 4.

$x = \dfrac{22}{4} = \dfrac{11}{2}$

The solution set is $\{\dfrac{11}{2}\}$.

25. $-7x+3 = -7$

$-7x+3-3 = -7-3$ Subtract 3 from both sides.

$-7x = -10$

$\dfrac{-7x}{-7} = \dfrac{-10}{-7}$ Divide both sides by -7.

$x = \dfrac{10}{7}$ The solution set is $\{\dfrac{10}{7}\}$.

29. $-16-4x = 9$

$-16+4x+16 = 9+16$ Add 16 to both sides.

$-4x = 25$

$\dfrac{-4x}{-4} = \dfrac{25}{-4}$ Divide both sides by -4.

$x = -\dfrac{25}{4}$

The solution set is $\{-\dfrac{25}{4}\}$.

33. $14y+15 = -33$

$14y+15-15 = -33-15$ Subtract 15 from both sides.

$14y = -48$

$\dfrac{14y}{14} = \dfrac{-48}{14}$ Divide both sides by 14.

$y = -\dfrac{48}{14} = -\dfrac{24}{7}$

The solution set is $\{-\dfrac{24}{7}\}$.

37. $17x-41 = -37$

$17x-41+41 = -37+41$ Add 41 to both sides.

$17x = 4$

$\dfrac{17x}{17} = \dfrac{4}{17}$ Divide both sides by 17.

$x = \dfrac{4}{17}$

The solution set is $\{\dfrac{4}{17}\}$.

41. Let n represent the number.

$$n+12 = 21$$
$$n = 9$$

45. Let c represent the cost of the item.

$$c+25 = 43$$
$$c = 18$$

The cost of the item is $18.

49. Let h represent his hourly rate. Then the product of the number of hours times the hourly rate equals total amount earned.

$$6h = 39$$
$$h = \frac{39}{6} = 6.5$$

His hourly rate was $6.50.

53. Let n represent the number.

$$19 = 3n+4$$
$$15 = 3n$$
$$5 = n$$

57. Let n represent the number.

$$6n-1 = 47$$
$$6n = 48$$
$$n = 8$$

61. Let c represent the cost of the ring.

$$550 = 2c-50$$
$$600 = 2c$$
$$300 = c$$

The cost of the ring was $300.

65. Let n represent the number of cars sold during December of 1983.

$$32 = 2n+4$$
$$28 = 2n$$
$$14 = n$$

They sold 14 cars during December of 1983.

Problem Set 3.3

1. $$2x+7+3x = 32$$
$$5x+7 = 32$$
$$5x+7-7 = 32-7$$
$$5x = 25$$
$$\frac{5x}{5} = \frac{25}{5}$$
$$x = 5$$

The solution set is $\{5\}$.

5. $$3y-1+2y-3 = 4$$
$$5y-4 = 4$$
$$5y-4+4 = 4+4$$
$$5y = 8$$
$$\frac{5y}{5} = \frac{8}{5}$$
$$y = \frac{8}{5}$$

The solution set is $\{\frac{8}{5}\}$.

9. $$-2n+1-3n+n-4 = 7$$
$$-4n-3 = 7$$
$$-4n-3+3 = 7+3$$
$$-4n = 10$$
$$\frac{-4n}{-4} = \frac{10}{-4}$$
$$n = -\frac{10}{4} = -\frac{5}{2}$$

The solution set is $\{-\frac{5}{2}\}$.

13. $$5x-7 = 6x-9$$
$$5x-7-5x = 6x-9-5x$$
$$-7 = x-9$$
$$-7+9 = x-9+9$$
$$2 = x$$

The solution set is $\{2\}$.

17. $$7y-3 = 5y+10$$
$$7y-3-5y = 5y+10-5y$$
$$2y-3 = 10$$
$$2y-3+3 = 10+3$$
$$2y = 13$$
$$\frac{2y}{2} = \frac{13}{2}$$
$$y = \frac{13}{2}$$

The solution set is $\{\frac{13}{2}\}$.

21. $$-2x-7 = -3x+10$$
$$-2x-7+3x = -3x+10+3x$$
$$x-7 = 10$$
$$x-7+7 = 10+7$$
$$x = 17$$

The solution set is $\{17\}$.

25.
$$
\begin{aligned}
-7-6x &= 9-9x \\
-7-6x+9x &= 9-9x+9x \\
-7+3x &= 9 \\
-7+3x+7 &= 9+7 \\
3x &= 16 \\
\frac{3x}{3} &= \frac{16}{3} \\
x &= \frac{16}{3}
\end{aligned}
$$

The solution set is $\{\frac{16}{3}\}$.

29.
$$
\begin{aligned}
5n-4-n &= -3n-6+n \\
4n-4 &= -2n-6 \\
4n-4+2n &= -2n-6+2n \\
6n-4 &= -6 \\
6n-4+4 &= -6+4 \\
6n &= -2 \\
\frac{6n}{6} &= \frac{-2}{6} \\
n &= -\frac{2}{6} = -\frac{1}{3}
\end{aligned}
$$

The solution set is $\{-\frac{1}{3}\}$.

33. Let n represent the number. Then 4n represents four times the number.

$$
\begin{aligned}
n+4n &= 85 \\
5n &= 85 \\
n &= 17
\end{aligned}
$$

37. Let n, n+2, and n+4 represent the three consecutive even numbers.

$$
\begin{aligned}
n+(n+2)+(n+4) &= 114 \\
3n+6 &= 114 \\
3n &= 108 \\
n &= 36
\end{aligned}
$$

The numbers are 36, 38, and 40.

41. Let n represent the number.

$$
\begin{aligned}
n+5n &= 3n-18 \\
6n &= 3n-18 \\
3n &= -18 \\
n &= -6
\end{aligned}
$$

45. Let a represent the smaller angle. Then 3a-20 represents the larger angle. Since they are supplementary angles, the sum of their measures is 180°.

$$
\begin{aligned}
a+(3a-20) &= 180 \\
4a-20 &= 180 \\
4a &= 200 \\
a &= 50
\end{aligned}
$$

The measures of the angles are 50° and 3(50)-20 = 130°.

49. Let x represent the price per share of the stock.

$$
\begin{aligned}
2x-17 &= 35 \\
2x &= 52 \\
x &= 26
\end{aligned}
$$

He paid $26 per share for the stock.

53. Let m represent the number of males; then 3m represents the number of females.

$$
\begin{aligned}
m+3m &= 600 \\
4m &= 600 \\
m &= 150
\end{aligned}
$$

Therefore, 150 males and 3(150) = 450 females attended the concert.

57. Let x represent the length of the shorter piece; then x+8 represents the length of the other piece.

$$
\begin{aligned}
x+(x+8) &= 20 \\
2x+8 &= 20 \\
2x &= 12 \\
x &= 6
\end{aligned}
$$

Problem Set 3.4

1. $7(x+2) = 21$

           $7x+14 = 21$ Apply distributive property to left side.

              $7x = 7$ Subtract 14 from both sides.

                $x = 1$ Divide both sides by 7.

The solution set is $\{1\}$.

> You may choose to divide both sides of the original equation by 7 producing $x+2 = 3$ and then complete the solution.

5. $-3(x+5) = 12$

         $-3x-15 = 12$ Apply distributive property to left side.

            $-3x = 27$ Add 15 to both sides.

              $x = -9$ Divide both sides by -3.

The solution set is $\{-9\}$.

9. $6(n+7) = 8$

         $6n+42 = 8$ Apply distributive property to left side.

            $6n = -34$ Subtract 42 from both sides.

$$n = -\frac{34}{6}\quad\text{Divide both sides by 6.}$$

$$n = -\frac{17}{3}\quad\text{Reduce.}$$

The solution set is $\{-\frac{17}{3}\}$.

13. $5(x-4) = 4(x+6)$

       $5x-20 = 4x+24$ Apply distributive property on both sides.

         $x-20 = 24$ Subtract $4x$ from both sides.

             $x = 44$

The solution set is $\{44\}$.

> We will discontinue giving reasons for each step but will continue to show enough of the work so that you can follow the steps. If a new technique is introduced, then we will indicate some of the reasons again.

17. $8(t+5) = 6(t-6)$

      $8t+40 = 6t-36$

      $2t+40 = -36$

         $2t = -76$

           $t = -38$

The solution set is $\{-38\}$.

21. $-2(x-6) = -(x-9)$ $\left(\begin{array}{l}-(x-9)\text{ means}\\-1(x-9).\end{array}\right)$

      $-2x+12 = -x+9$

        $-x+12 = 9$

          $-x = -3$

            $x = 3$

The solution set is $\{3\}$.

25. $3(n-10)-5(n+12) = -86$

      $3n-30-5n-60 = -86$

          $-2n-90 = -86$

             $-2n = 4$

               $n = -2$

The solution set is $\{-2\}$.

29. $-(x+2)+2(x-3) = -2(x-7)$

      $-x-2+2x-6 = -2x+14$

            $x-8 = -2x+14$

         $3x-8 = 14$

            $3x = 22$

$$x = \frac{22}{3}$$

The solution set is $\{\frac{22}{3}\}$.

33.
$$-(a-1)-(3a-2) = 6+2(a-1)$$
$$-a+1-3a+2 = 6+2a-2$$
$$-4a+3 = 2a+4$$
$$-6a+3 = 4$$
$$-6a = 1$$
$$a = -\frac{1}{6}$$

The solution set is $\{-\frac{1}{6}\}$.

37.
$$3-7(x-1) = 9-6(2x+1)$$
$$3-7x+7 = 9-12x-6$$
$$-7x+10 = -12x+3$$
$$5x+10 = 3$$
$$5x = -7$$
$$x = -\frac{7}{5}$$

The solution set is $\{-\frac{7}{5}\}$.

> For Problems 39-60, we begin each solution by multiplying both sides of the given equation by the least common denominator of all of the denominators in the equation. This has the effect of "clearing the equation of all fractions."

41.
$$\frac{5}{6}x+\frac{1}{4} = -\frac{9}{4}$$
$$12(\frac{5}{6}x+\frac{1}{4}) = 12(-\frac{9}{4})$$
$$10x+3 = -27$$
$$10x = -30$$
$$x = -3$$

The solution set is $\{-3\}$.

45.
$$\frac{n}{3}+\frac{5n}{6} = \frac{1}{8}$$
$$24(\frac{n}{3}+\frac{5n}{6}) = 24(\frac{1}{8})$$
$$8n+20n = 3$$
$$28n = 3$$
$$n = \frac{3}{28}$$

The solution set is $\{\frac{3}{28}\}$.

49.
$$\frac{h}{6}+\frac{h}{8} = 1$$
$$24(\frac{h}{6}+\frac{h}{8}) = 24(1)$$
$$4h+3h = 24$$
$$7h = 24$$
$$h = \frac{24}{7}$$

The solution set is $\{\frac{24}{7}\}$.

53.
$$\frac{x-1}{5}-\frac{x+4}{6} = -\frac{13}{15}$$
$$30(\frac{x-1}{5}-\frac{x+4}{6}) = 30(-\frac{13}{15})$$
$$6(x-1)-5(x+4) = -26$$
$$6x-6-5x-20 = -26$$
$$x-26 = -26$$
$$x = 0$$

The solution set is $\{0\}$.

57.
$$\frac{x-2}{8}-1 = \frac{x+1}{4}$$
$$8(\frac{x-2}{8}-1) = 8(\frac{x+1}{4})$$
$$x-2-8 = 2(x+1)$$
$$x-10 = 2x+2$$
$$-12 = x$$

The solution set is $\{-12\}$.

61. Let n and n+1 represent the consecutive whole numbers.
$$n+4(n+1) = 39$$
$$n+4n+4 = 39$$
$$5n = 35$$
$$n = 7$$

The numbers are 7 and 8.

65. Let n represent the smaller number;
then 17-n represents the larger
number.
$$2n = (17-n)+1$$
$$2n = 17-n+1$$
$$2n = 18-n$$
$$3n = 18$$
$$n = 6$$

The numbers are 6 and 17-6 = 11.

69. Let n represent the smaller number; then n+6 represents the
larger number.
$$\frac{1}{2}(n+6) = \frac{1}{3}n+5$$
$$6[\frac{1}{2}(n+6)] = 6(\frac{1}{3}n+5)$$
$$3(n+6) = 2n+30$$
$$3n+18 = 2n+30$$
$$n = 12$$

The numbers are 12 and 12+6 = 18.

73. Let n represent the number of nickels, n+5 the number of dimes, and 3n+4 the number of quarters.

$$n+(n+5)+(3n+4) = 69$$
$$5n+9 = 69$$
$$5n = 60$$
$$n = 12$$

She has 12 nickels, 12+5 = 17 dimes, and 3(12)+4 = 40 quarters.

77. Let d represent the number of dimes and 18-d the number of quarters.

$$10d+25(18-d) = 330$$
$$10d+450-25d = 330$$
$$-15d = -120$$
$$d = 8$$

She has 8 dimes and 18-8 = 10 quarters.

81. Let a represent the measure of the angle. Then 90-a represents its complement and 180-a its supplement.

$$180-a = 2(90-a)+30$$
$$180-a = 180-2a+30$$
$$180-a = -2a+210$$
$$a = 30$$

The angle has a measure of 30°.

85. Let a represent the measure of the angle. Then 90-a represents its complement and 180-a its supplement.

$$180-a = 3(90-a)-10$$
$$180-a = 270-3a-10$$
$$180-a = -3a+260$$
$$2a = 80$$
$$a = 40$$

The angle has a measure of 40°.

Problem Set 3.5

1. $\dfrac{x}{6} = \dfrac{3}{2}$

$2x = 18$ Cross products are equal.

$x = 9$

The solution set is {9}.

5. $\dfrac{x}{3} = \dfrac{5}{2}$

$2x = 15$ Cross products are equal.

$x = \dfrac{15}{2}$

The solution set is $\{\dfrac{15}{2}\}$.

9. $\dfrac{x+1}{6} = \dfrac{x+2}{4}$

$6(x+2) = 4(x+1)$ Cross products are equal.

$6x+12 = 4x+4$

$2x = -8$

$x = -4$

The solution set is {-4}.

13. $\dfrac{-1}{x-7} = \dfrac{5}{x-1}$

$5(x-7) = -1(x-1)$ Cross products are equal.

$5x-35 = -x+1$

$6x = 36$

$x = 6$

The solution set is {6}.

17. $\dfrac{n+1}{n} = \dfrac{8}{7}$

    $8n = 7(n+1)$ Cross products
                      are equal.

    $8n = 7n+7$

     $n = 7$

The solution set is {7}.

21. $\dfrac{300-n}{n} = \dfrac{3}{2}$

    $3n = 2(300-n)$ Cross products
                        are equal.

    $3n = 600-2n$

    $5n = 600$

     $n = 120$

The solution set is {120}.

25. $\dfrac{11}{20} = \dfrac{n}{100}$

    $20n = 1100$

      $n = 55$

Therefore, $\dfrac{11}{20} = \dfrac{55}{100} = 55\%$.

29. $\dfrac{1}{6} = \dfrac{n}{100}$

    $6n = 100$

     $n = 16\dfrac{2}{3}$

Therefore, $\dfrac{1}{6} = \dfrac{16\frac{2}{3}}{100} = 16\dfrac{2}{3}\%$.

33. $\dfrac{3}{2} = \dfrac{n}{100}$

    $2n = 300$

     $n = 150$

Therefore, $\dfrac{3}{2} = \dfrac{150}{100} = 150\%$.

37. Let n represent the number.

    $.07(38) = n$

      $2.66 = n$

41. Let n represent the number that represents percent.

    $76 = 95n$

    $\dfrac{76}{95} = n$

    $.8 = n$

Therefore, 76 is 80% of 95.

45. Let n represent the number that represents percent.

    $46 = 40n$

    $\dfrac{46}{40} = n$

    $1.15 = n$

Therefore, 46 is 115% of 40.

49. Let ℓ and w represent the length and width of the room measured in feet.

    $\dfrac{1}{2\frac{1}{2}} = \dfrac{6}{w}$ $\dfrac{1}{3\frac{1}{4}} = \dfrac{6}{\ell}$

      $w = 6(2\tfrac{1}{2})$ $\ell = 6(3\tfrac{1}{4})$

      $w = 15$ $\ell = 19\tfrac{1}{2}$

The room measures 15 feet by $19\tfrac{1}{2}$ feet.

53. Let ℓ represent the length.

    $\dfrac{5}{2} = \dfrac{\ell}{24}$

    $2\ell = 120$

     $\ell = 60$

The length is 60 centimeters.

57. Let p represent the number of pounds needed for 2500 square feet.

    $\dfrac{20}{p} = \dfrac{1500}{2500}$

    $1500p = 20(2500)$

    $1500p = 50000$

        $p = 33\tfrac{1}{3}$

It will take $33\tfrac{1}{3}$ pounds of fertilizer to cover 2500 square feet of lawn.

61. Let ℓ represent the length of the rectangle. Then $25-\ell$ represents the width because length plus width equals one-half of the perimeter.

$$\frac{\ell}{25-\ell} = \frac{3}{2}$$
$$2\ell = 3(25-\ell)$$
$$2\ell = 75-3\ell$$
$$5\ell = 75$$
$$\ell = 15$$

The length is 15 inches and the width is $25-15 = 10$ inches.

65. Let x represent the amount that the child will receive. Then $180,000-x$ represents the amount that the cancer fund will receive.

$$\frac{x}{180000-x} = \frac{5}{1}$$
$$5(180,000-x) = x$$
$$900,000-5x = x$$
$$900,000 = 6x$$
$$150,000 = x$$

The child will receive $150,000.

Problem Set 3.6

1. $x - .36 = .75$
    $x - .36 + .36 = .75 + .36$ Add .36 to both sides.
    $x = 1.11$

    The solution set is $\{1.11\}$.

5. $.62 - y = .14$
    $.62 - y - .62 = .14 - .62$ Subtract .62 from both sides.
    $-y = -.48$
    $y = .48$

    The solution set is $\{.48\}$.

9. $x = 3.36 - .12x$
    $100(x) = 100(3.36 - .12x)$ Multiply both sides by 100.
    $100x = 336-12x$
    $112x = 336$
    $x = 3$

    The solution set is $\{3\}$.

13. $s = 42 + .4s$
    $10(s) = 10(42 + .4s)$ Multiply both sides by 10.
    $10 = 420+4s$
    $6s = 420$
    $s = 70$

    The solution set is $\{70\}$.

17. $.09x + .1(2x) = 130.5$
    $100(.09x + .1(2x)) = 100(130.5)$
    $9x+20x = 13050$
    $29x = 13050$
    $x = 450$

    The solution set is $\{450\}$.

21. $.09x = 550 - .11(5400-x)$
    $100(.09x) = 100[550 - .11(5400-x)]$
    $9x = 55000 - 11(5400-x)$
    $9x = 55000 - 59400 + 11x$
    $-2x = -4400$
    $x = 2200$

    The solution set is $\{2200\}$.

25. Let c represent the cost of the sweater.
    $c = 48 - .25(48)$
    $c = 48 - 12$
    $c = 36$

    The cost of the sweater is $36.

29. Let r represent the rate of discount. His discount was $180-$126 = $54. Therefore, the rate of discount was

$$r = \frac{54}{180} = \frac{3}{10} = \frac{30}{100} = 30\%.$$

33. Let s represent the selling price. The guideline "selling price equals cost plus profit" can be used to set up the equation.

$$s = 3 + .55(3)$$
$$s = 4.65$$

He should sell them for $4.65 each.

37. Let r represent the rate of profit. The profit is $44.80−$32 = $12.80. Therefore, as a percent of the cost, the profit is

$$r = \frac{12.80}{32} = .4 = 40\%.$$

41. Let x be the amount invested at 9% and 1200−x the amount at 12%.

$$.09x + .12(1200-x) = 129$$
$$9x+12(1200-x) = 12900$$
$$9x+14400-12x = 12900$$
$$-3x = -1500$$
$$x = 500$$

Therefore, $500 is invested at 9% and $1200−$500 = $700 at 12%.

45. Let x be the amount invested at 10% and 2300−x the amount at 12%.

$$.12(2300-x) = .10x+100$$
$$12(2300-x) = 10x+10000$$
$$27600-12x = 10x+10000$$
$$17600 = 22x$$
$$800 = x$$

Therefore, $800 is invested at 10% and $2300−$800 = $1500 at 12%.

Chapter 4

Problem Set 4.1

1. $d = rt$
    $336 = 48t$
    $7 = t$

5. $F = \frac{9}{5}C + 32$

    $68 = \frac{9}{5}C + 32$

    $36 = \frac{9}{5}C$

    $\frac{5}{9}(36) = \frac{5}{9}(\frac{9}{5}C)$

    $20 = C$

9. $A = P + Prt$
    $652 = 400 + 400(.07)t$
    $252 = 28t$
    $9 = t$

13. Let w represent its width. We can use the guideline "length plus width equals one-half of the perimeter." First, we must change $3\frac{1}{4}$ feet to 39 inches.

    $39 + w = \frac{1}{2}(108)$

    $39 + w = 54$
    $w = 15$

    Its width is 15 inches.

17. Each piece of wood has an area of $(30)(60) = 1800$ square centimeters. There are 50 pieces, so we have a total area of $50(1800) = 90000$ square centimeters. Now we can change to square meters by dividing by 10,000. Therefore, 90000 square centimeters equals 9 square meters. We need one liter of paint at a cost of \$2.

21. 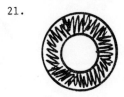 The diameter of the large circle is 4 centimeters and the diameter of the small circle is 2 centimeters. The area of the shaded region is the area of the larger circle minus the area of the smaller circle.

    $A = \Pi(2)^2 - \Pi(1)^2 = 4\Pi - \Pi = 3\Pi$

    There are 50 washers, so we have $50(3\Pi) = 150\Pi$ square centimeters of metal.

25. $S = 4\Pi r^2$

    $S = 4\Pi(9)^2$

    $S = 324\Pi$ square inches

    $V = \frac{4}{3}\Pi r^3$

    $V = \frac{4}{3}\Pi(9)^3$

    $V = 972\Pi$ cubic inches

29. $V = \frac{1}{3}\Pi r^2 h$

    $324\Pi = \frac{1}{3}\Pi(9)^2 h$

    $324\Pi = \frac{81\Pi}{3}h$

    $324\Pi = 27\Pi h$
    $12 = h$

33. $V = Bh$

    $\frac{V}{B} = \frac{Bh}{B}$ Divide both sides by B.

    $\frac{V}{B} = h$

37.　　　　$P = 2\ell+2w$

　　　　$P-2\ell = 2w$　　　　Subtract 2 from both sides.

　　　　$\dfrac{P-2\ell}{2} = w$　　　Divide both sides by 2.

41.　　　　$F = \dfrac{9}{5}C + 32$

　　　　$F-32 = \dfrac{9}{5}C$　　　Subtract 32 from both sides.

　　　　$\dfrac{5}{9}(F-32) = \dfrac{5}{9}(\dfrac{9}{5}C)$　　Multiply both sides by $\dfrac{5}{9}$.

　　　　$\dfrac{5}{9}(F-32) = C$

45.　　　$3x+7y = 9$

　　　　　$3x = 9-7y$　　　Subtract 7y from both sides.

　　　　　$x = \dfrac{9-7y}{3}$　　　Divide both sides by 3.

49.　　　$-2x+11y = 14$

　　　　　$11y = 14+2x$　　　Add 2x to both sides.

　　　$11y-14 = 2x$　　　Subtract 14 from both sides.

　　　$\dfrac{11y-14}{2} = x$　　　Divide both sides by 2.

53.　　$\dfrac{x-2}{4} = \dfrac{y-3}{6}$

　　$4(y-3) = 6(x-2)$　　　Cross products are equal.

　　$4y-12 = 6x-12$

　　　$4y = 6x$　　　Add 12 to both sides.

　　　$y = \dfrac{6x}{4} = \dfrac{3x}{2}$

57.　　$\dfrac{x+6}{2} = \dfrac{y+4}{5}$

　　$5(x+6) = 2(y+4)$　　　Cross products are equal.

　　$5x+30 = 2y+8$

　　　$5x = 2y-22$　　　Subtract 30 from both sides.

　　　$x = \dfrac{2y-22}{5}$　　　Divide both sides by 5.

Problem Set 4.2

1.　　　$i = Prt$

　　　$750 = 750(.08)t$

　　　$1 = .08t$

　　　$100 = 8t$

　　　$\dfrac{100}{8} = t$

　　　$12\dfrac{1}{2} = t$

　　It will take $12\dfrac{1}{2}$ years.

5.

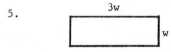

The length of a rectangle plus its width equals one-half of the perimeter.

　　　$3w+w = 56$

　　　$4w = 56$

　　　$w = 14$

The width is 14 inches and the length is $3(14) = 42$ inches.

9.

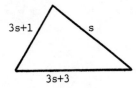

$\ell + \frac{1}{2}\ell - 3 = \frac{1}{2}(42)$

$\frac{3}{2}\ell - 3 = 21$

$\frac{3}{2}\ell = 24$

$\ell = 16$

The length is 16 inches and the width is $\frac{1}{2}(16)-3 = 5$ inches. Therefore, the area is 5(16) = 80 square inches.

13.

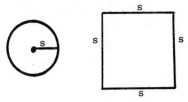

Let s represent the length of the "first" side.

$s+(3s+1)+(3s+3) = 46$

$7s+4 = 46$

$7s = 42$

$s = 6$

The sides are of length 6 centimeters, 3(6)+1 = 19 centimeters, and 3(6)+3 = 21 centimeters.

17.

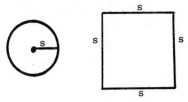

$2\pi s = 4s + 15.96$

$2\pi s - 4s = 15.96$

$s(2\pi - 4) = 15.96$

$s = \frac{15.96}{2\pi - 4} = \frac{15.96}{2(3.14)-4} = \frac{15.96}{2.28} = 7$

A radius of the circle is 7 centimeters long.

21.

325 miles

Let t represent the time of the freight train. Therefore, t also represents the time of the passenger train since they start and stop at the same time.

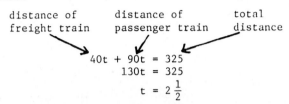

$$40t + 90t = 325$$
$$130t = 325$$
$$t = 2\frac{1}{2}$$

They will both travel for $2\frac{1}{2}$ hours.

25.

	time	rate	distance
east-bound	$9\frac{1}{2}$	r+8	$9\frac{1}{2}(r+8)$
west-bound	$9\frac{1}{2}$	r	$9\frac{1}{2}(r)$

Since the total distance was 1292 miles, we can set up and solve the following equation.

$$\frac{19}{2}(r+8) + \frac{19}{2}r = 1292$$
$$19(r+8) + 19r = 2584$$
$$19r + 152 + 19r = 2584$$
$$38r = 2432$$
$$r = 64$$

The west-bound train travels 64 miles per hour and the east-bound train 72 miles per hour.

Problem Set 4.3

1. Let x be the amount of pure acid to be added.

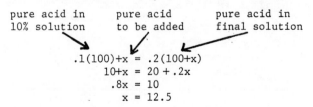

$$.1(100) + x = .2(100 + x)$$
$$10 + x = 20 + .2x$$
$$.8x = 10$$
$$x = 12.5$$

We must add 12.5 milliliters of pure acid.

5. Let x be the amount of 30% solution and 10-x be the amount of 50% solution.

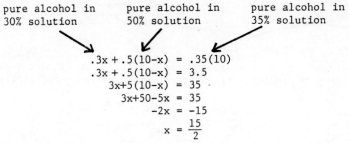

$$
\begin{aligned}
.3x + .5(10-x) &= .35(10) \\
.3x + .5(10-x) &= 3.5 \\
3x + 5(10-x) &= 35 \\
3x + 50 - 5x &= 35 \\
-2x &= -15 \\
x &= \frac{15}{2}
\end{aligned}
$$

We must mix $7\frac{1}{2}$ quarts of the 30% solution with $10 - 7\frac{1}{2} = 2\frac{1}{2}$ quarts of the 50% solution.

9. Let x be the amount of mixture drained out and also the amount of pure antifreeze to be added.

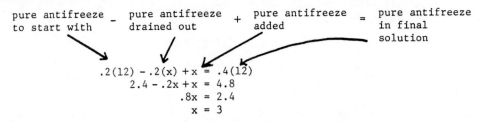

$$
\begin{aligned}
.2(12) - .2(x) + x &= .4(12) \\
2.4 - .2x + x &= 4.8 \\
.8x &= 2.4 \\
x &= 3
\end{aligned}
$$

We must drain out 3 quarts of the 20% solution and add 3 quarts of pure antifreeze.

13. $.1(30) = 3$ and $.2(50) = 10$

The final mixture will contain $30+50 = 80$ ounces, of which there are $3+10 = 13$ ounces of grapefruit juice. Therefore, letting x represent the percent of grapefruit juice we obtain

$$
x = \frac{13}{80} = .1625 = 16.25\%.
$$

17.

	rate	time	distance
Butch	2	t	$2t$
Dick	$3\frac{1}{2}$	$t - \frac{1}{2}$	$3\frac{1}{2}(t - \frac{1}{2})$

Since Dick is to catch Butch, they walk the same distance.

$$
\begin{aligned}
2t &= \frac{7}{2}(t - \frac{1}{2}) \\
4t &= 7(t - \frac{1}{2}) \\
4t &= 7t - \frac{7}{2} \\
-3t &= -\frac{7}{2} \\
t &= 1\frac{1}{6}
\end{aligned}
$$

Therefore, it will take Dick $1\frac{1}{6} - \frac{1}{2} = \frac{7}{6} - \frac{3}{6} = \frac{4}{6} = \frac{2}{3}$ of an hour (40 minutes) to catch Butch.

21. We can represent the various ages as follows:

$$x : \text{Abby's present age}$$
$$x+21 : \text{her mother's present age}$$
$$x+10 : \text{Abby's age in ten years}$$
$$x+31 : \text{her mother's age in ten years}$$

$$x+31 = 2(x+10)-3$$
$$x+31 = 2x+20-3$$
$$x+31 = 2x+17$$
$$14 = x$$

Abby's present age is 14 and her mother's age is 14+21 = 35.

25. Let t represent the time of Steve's trip. Then $t - \frac{1}{2}$ represents his time if he had increased his speed.

$$14t = 16(t - \frac{1}{2})$$
$$14t = 16t-8$$
$$-2t = -8$$
$$t = 4$$

Steve rode for 4 hours at 14 miles per hour; therefore, this distance was 4(14) = 56 miles.

Problem Set 4.4

1. The left side simplifies to 2(3)-4(5) = 6-20 = -14 and the right side to 5(3)-2(-1)+4 = 15+2+4 = 21. Since -14 < 21, the given inequality is true.

5. The left side simplifies to $(-\frac{1}{2})(\frac{4}{9}) = -\frac{2}{9}$ and the right side to $(\frac{3}{5})(-\frac{1}{3}) = -\frac{1}{5}$. Since $-\frac{2}{9} < -\frac{1}{5}$, the given inequality is false.

9. The left side simplifies to .16 + .34 = .5 and the right side to .23 + .17 = .4. Since .5 > 4, the given inequality is true.

25.
$$x-4 \geq -13$$
$$x-4+4 \geq -13+4 \qquad \text{Add 4 to both sides.}$$
$$x \geq -9$$

The solution set is $\{x \mid x \geq -9\}$.

29.
$$6x < 20$$
$$\frac{6x}{6} < \frac{20}{6} \qquad \text{Divide both sides by 6.}$$
$$x < \frac{20}{6}$$
$$x < \frac{10}{3}$$

The solution set is $\{x \mid x < \frac{10}{3}\}$.

33. $-7n \leq -56$

$\dfrac{-7n}{-7} \geq \dfrac{-56}{-7}$ Divide both sides by -7 which reverses the inequality.

$n \geq 8$

The solution set is $\{n \mid n \geq 8\}$.

37. $16 < 9+n$

$16-9 < 9+n-9$ Subtract 9 from both sides.

$7 < n$

$n > 7$ $7 < n$ means $n > 7$.

The solution set is $\{n \mid n > 7\}$.

41. $4x-3 \leq 21$

$4x-3+3 \leq 21+3$ Add 3 to both sides.

$4x \leq 24$

$\dfrac{4x}{4} \leq \dfrac{24}{4}$ Divide both sides by 4.

$x \leq 6$

The solution set is $\{x \mid x \leq 6\}$.

45. $6x+2 < 18$

$6x+2-2 < 18-2$ Subtract 2 from both sides.

$6x < 16$

$\dfrac{6x}{6} < \dfrac{16}{6}$ Divide both sides by 6.

$x < \dfrac{16}{6}$

$x < \dfrac{8}{3}$

The solution set is $\{x \mid x < \dfrac{8}{3}\}$.

49. $-2 < -3x+1$

$-2-1 < -3x$ Subtract 1 from both sides.

$-3 < -3x$

$\dfrac{-3}{-3} > \dfrac{-3x}{-3}$ Divide both sides by -3 which reverses the inequality.

$1 > x$

$x < 1$ $1 > x$ means $x < 1$.

The solution set is $\{x \mid x < 1\}$.

53. $5x-4-3x > 24$

$\quad\quad 2x-4 > 24$ Combine similar terms on the left side.

$\quad 2x-4+4 > 24+4$ Add 4 to both sides.

$\quad\quad\quad 2x > 28$

$\quad\quad\quad \dfrac{2x}{2} > \dfrac{28}{2}$ Divide both sides by 2.

$\quad\quad\quad\quad x > 14$

The solution set is $\{x \mid x > 14\}$.

57. $-5 \geq 3t-4-7t$

$\quad\quad -5 \geq -4t-4$ Combine similar terms.

$\quad -5+4 \geq -4t-4+4$ Add 4 to both sides.

$\quad\quad -1 \geq -4t$

$\quad\quad \dfrac{-1}{-4} \leq \dfrac{-4t}{-4}$ Divide both sides by -4 which reverses the inequality.

$\quad\quad \dfrac{1}{4} \leq t$

$\quad\quad t \geq \dfrac{1}{4}$ $\dfrac{1}{4} \leq t$ means $t \geq \dfrac{1}{4}$.

The solution set is $\{t \mid t \geq \dfrac{1}{4}\}$.

Problem Set 4.5

1. $3x+4 > x+8$

$\quad 3x+4-x > x+8-x$ Subtract x from both sides.

$\quad\quad 2x+4 > 8$

$\quad\quad 2x+4-4 > 8-4$ Subtract 4 from both sides.

$\quad\quad\quad 2x > 4$

$\quad\quad\quad \dfrac{2x}{2} > \dfrac{4}{2}$ Divide both sides by 2.

$\quad\quad\quad\quad x > 2$

The solution set is $\{x \mid x > 2\}$.

5. $6x+7 > 3x-3$

$\quad 6x+7-3x > 3x-3-3x$ Subtract 3x from both sides.

$\quad\quad 3x+7 > -3$

$\quad 3x+7-7 > -3-7$ Subtract 7 from both sides.

$\quad\quad\quad 3x > -10$

$\quad\quad\quad \dfrac{3x}{3} > \dfrac{-10}{3}$ Divide both sides by 3.

$\quad\quad\quad x > -\dfrac{10}{3}$

The solution set is $\{x \mid x > -\dfrac{10}{3}\}$.

9. $2t+9 \geq 4t-13$

$\quad 2t+9-4t \geq 4t-13-4t$ Subtract 4t from both sides.

$\quad\quad -2t+9 \geq -13$

$\quad -2t+9-9 \geq -13-9$ Subtract 9 from both sides.

$\quad\quad -2t \geq -22$

$\quad\quad \dfrac{-2t}{-2} \leq \dfrac{-22}{-2}$ Divide both sides by -2 which reverses the inequality.

$\quad\quad t \leq 11$

The solution set is $\{t \mid t \leq 11\}$.

29

13. $-4x+6 > -2x+1$

    $-4x+6+2x > -2x+1+2x$ Add 2x to both sides.

        $-2x+6 > 1$

    $-2x+6-6 > 1-6$ Subtract 6 from both sides.

        $-2x > -5$

       $\dfrac{-2x}{-2} < \dfrac{-5}{-2}$ Divide both sides by -2 which reverses the inequality.

        $x < \dfrac{5}{2}$

    The solution set is $\{x \mid x < \frac{5}{2}\}$.

17. $2(n+3) > 9$

    $2n+6 > 9$ Apply distributive property to left side.

    $2n+6-6 > 9-6$ Subtract 6 from both sides.

    $2n > 3$

    $\dfrac{2n}{2} > \dfrac{3}{2}$ Divide both sides by 2.

    $n > \dfrac{3}{2}$

    The solution set is $\{n \mid n > \frac{3}{2}\}$.

21. $-2(x+6) > -17$

    $-2x-12 > -17$ Apply distributive property to left side.

    $-2x-12+12 > -17+12$ Add 12 to both sides.

    $-2x > -5$

    $\dfrac{-2x}{-2} < \dfrac{-5}{-2}$ Divide both sides by -2 which reverses the inequality.

    $x < \dfrac{5}{2}$

    The solution set is $\{x \mid x < \frac{5}{2}\}$.

25. $4(x+3) > 6(x-5)$

    $4x+12 > 6x-30$ Apply distributive property to both sides.

    $4x+12-6x > 6x-30-6x$ Subtract 6x from both sides.

    $-2x+12 > -30$

    $-2x+12-12 > -30-12$ Subtract 12 from both sides.

    $-2x > -42$

    $\dfrac{-2x}{-2} < \dfrac{-42}{-2}$ Divide both sides by -2 which reverses the inequality.

    $x < 21$

    The solution set is $\{x \mid x < 21\}$.

29. $5(n+1)-3(n-1) > 9$

    $5n+5-3n+3 > -9$ Apply the distributive property twice on the left side.

    $2n+8 > -9$ Combine similar terms.

    $2n+8-8 > -9-8$ Subtract 8 from both sides.

    $2n > -17$

    $\dfrac{2n}{2} > \dfrac{-17}{2}$ Divide both sides by 2.

    $n > -\dfrac{17}{2}$

    The solution set is $\{n \mid n > -\frac{17}{2}\}$.

33.

$$\frac{3}{4}n - \frac{5}{6}n < \frac{3}{8}$$

$$24\left(\frac{3}{4}n - \frac{5}{6}n\right) < 24\left(\frac{3}{8}\right) \qquad \text{Multiply both sides by 24.}$$

$$18n - 20n < 9$$

$$-2n < 9$$

$$\frac{-2n}{-2} > \frac{9}{-2} \qquad \text{Divide both sides by } -2 \text{ which reverses the inequality.}$$

$$n > -\frac{9}{2}$$

The solution set is $\{n \mid n > -\frac{9}{2}\}$.

37.

$$n \geq 3.4 + .15n$$

$$100(n) \geq 100(3.4 + .15n) \qquad \text{Multiply both sides by 100.}$$

$$100n \geq 340 + 15n$$

$$85n \geq 340$$

$$n \geq 4$$

The solution set is $\{n \mid n \geq 4\}$.

41.

$$.06x + .08(250-x) \geq 19$$

$$100[.06x + .08(250-x)] \geq 100(19) \qquad \text{Multiply both sides by 100.}$$

$$6x + 8(250-x) \geq 1900$$

$$6x + 2000 - 8x \geq 1900$$

$$-2x \geq -100$$

$$x \leq 50$$

The solution set is $\{x \mid x \leq 50\}$.

45.

$$\frac{x+2}{6} - \frac{x+1}{5} < -2$$

$$30\left(\frac{x+2}{6} - \frac{x+1}{5}\right) < 30(-2) \qquad \text{Multiply both sides by 30.}$$

$$5(x+2) - 6(x+1) < -60$$

$$5x + 10 - 6x - 6 < -60$$

$$-x + 4 < -60$$

$$-x < -64$$

$$x > 64$$

The solution set is $\{x \mid x > 64\}$.

49.

$$\frac{x-3}{7} - \frac{x-2}{4} \leq \frac{9}{14}$$

$$28\left(\frac{x-3}{7} - \frac{x-2}{4}\right) \leq 28\left(\frac{9}{14}\right) \qquad \text{Multiply both sides by 28.}$$

$$4(x-3) - 7(x-2) \leq 18$$

$$4x - 12 - 7x + 14 \leq 18$$

$$-3x + 2 \leq 18$$

$$-3x \leq 16$$

$$x \geq -\frac{16}{3}$$

The solution set is $\{x \mid x \geq -\frac{16}{3}\}$.

53. Since it is an "or" statement, the solution set consists of all numbers less than -2 along with all numbers greater than 1.

57. Since it is an "and" statement we are looking for all numbers that satisfy both inequalities at the same time. Thus, any number greater than 2 will work.

61. Since it is an "and" statement we are looking for all numbers that satisfy both inequalities at the same time. There are no numbers that are both greater than 3 and less than -1. So the solution set is $\emptyset$.

65. Since it is an "or" statement, we want all numbers greater than -4 along with all numbers less than 3. Thus, the solution set is the entire set of real numbers.

69. Let w represent the width of the rectangle. Also remember that "length plus width equals one-half of the perimeter" of a rectangle.

$$w+20 \leq 35$$
$$w \leq 15$$

Thus, 15 inches is the largest possible value for the width.

73. Let x be his average on the last two exams.

$$\frac{96+90+94+2x}{5} > 92$$
$$280+2x > 460$$
$$2x > 180$$
$$x > 90$$

He must have an average greater than 90 on the last two exams.

77. Let x be his score on the final round.

$$\frac{82+84+78+79+x}{5} \leq 80$$
$$323+x \leq 400$$
$$x \leq 77$$

He must shoot 77 or less.

Problem Set 4.6

1. $|x| = 4$ is equivalent to $x = -4$ or $x = 4$. Therefore, the solution set is $\{-4,4\}$.

5. $|x| \geq 2$ means that x must be equal to or more than 2 units away from zero. Therefore, $|x| \geq 2$ is equivalent to $x \leq -2$ or $x \geq 2$. The solution set is $\{x \mid x \leq -2$ or $x \geq 2\}$.

9. $|x-1| = 2$ is equivalent to $x-1 = -2$ or $x-1 = 2$.

$$x-1 = -2 \text{ or } x-1 = 2$$
$$x = -1 \text{ or } \quad x = 3$$

Ths solution set is $\{-1,3\}$.

13. $|x+1| > 3$ is equivalent to $x+1 < -3$ or $x+1 > 3$.

$$x+1 < -3 \text{ or } x+1 > 3$$
$$x < -4 \text{ or } \quad x > 2$$

The solution set is $\{x \mid x < -4$ or $x > 2\}$.

17. $|5x-2| = 4$ is equivalent to $5x-2 = -4$ or $5x-2 = 4$.

$$5x-2 = -4 \text{ or } 5x-2 = 4$$
$$5x = -2 \text{ or } \quad 5x = 6$$
$$x = -\frac{2}{5} \text{ or } \quad x = \frac{6}{5}$$

The solution set is $\{-\frac{2}{5}, \frac{6}{5}\}$.

21. $|4x+3| < 2$ is equivalent to $4x+3 > -2$ and $4x+3 < 2$.

$$4x+3 > -2 \text{ and } 4x+3 < 2$$
$$4x > -5 \text{ and } \quad 4x < -1$$
$$x > -\frac{5}{4} \text{ and } \quad x < -\frac{1}{4}$$

The solution set is $\{x \mid x > -\frac{5}{4} \text{ and } x < -\frac{1}{4}\}$.

25. Solve $3x-2 = 0$.

$$3x-2 = 0$$
$$3x = 2$$
$$x = \frac{2}{3}$$

The solution set is $\{x \mid x \neq \frac{2}{3}\}$.

29. $|2x+1| > 9$ is equivalent to $2x+1 < -9$ or $2x+1 > 9$.

$$2x+1 < -9 \text{ or } 2x+1 > 9$$
$$2x < -10 \text{ or } \quad 2x > 8$$
$$x < -5 \text{ or } \quad x > 4$$

The solution set is $\{x \mid x < -5 \text{ or } x > 4\}$.

33. $|-3x-1| = 17$ is equivalent to $-3x-1 = -17$ or $-3x-1 = 17$.

$$-3x-1 = -17 \text{ or } -3x-1 = 17$$
$$-3x = -16 \text{ or } \quad -3x = 18$$
$$x = \frac{16}{3} \text{ or } \quad x = -6$$

The solution set is $\{-6, \frac{16}{3}\}$.

37. $|5x+3| \geq 18$ is equivalent to $5x+3 \leq -18$ or $5x+3 \geq 18$.

$$5x+3 \leq -18 \text{ or } 5x+3 \geq 18$$
$$5x \leq -21 \text{ or } \quad 5x \geq 15$$
$$x \leq -\frac{21}{5} \text{ or } \quad x \geq 3$$

The solution set is $\{x \mid x \leq -\frac{21}{5} \text{ or } x \geq 3\}$.

41. $|-2x+1| > 6$ is equivalent to $-2x+1 < -6$ or $-2x+1 > 6$.

$$-2x+1 < -6 \text{ or } -2x+1 > 6$$
$$-2x < -7 \text{ or } \quad -2x > 5$$
$$x > \frac{7}{2} \text{ or } \quad x < -\frac{5}{2}$$

The solution set is $\{x \mid x < -\frac{5}{2} \text{ or } x > \frac{7}{2}\}$.

45. Since the absolute value of any number is nonnegative, the statement $|x-6| > -4$ will be true for all real numbers. The solution set is $\{x \mid x \text{ is a real number}\}$.

49. Since the absolute value of any number is nonnegative, the statement $|x+6| \leq 0$ will only hold if $x = -6$. The solution set is $\{-6\}$.

Chapter 5

Problem Set 5.1

1. The degree of $7x^2y+6xy$ is 3 because the degree of the term $7x^2y$ is 3.

5. The degree of $5x^3-x^2-x+3$ is 3 because the degree of $5x^3$ is 3.

9. $(3x+4)+(5x+7) = (3+5)x+(4+7) = 8x+11$

13. $(-2x^2+7x-9)+(4x^2-9x-14) = (-2+4)x^2+(7-9)x+(-9-14)$
$$= 2x^2-2x-23$$

17. $(2x^2-x+4)+(-5x^2-7x-2)+(9x^2+3x-6) = (2-5+9)x^2+(-1-7+3)x+(4-2-6)$
$$= 6x^2-5x-4$$

21. $(3x-7)-(5x-2) = 3x-7-5x+2 = (3-5)x-7+2$
$$= -2x-5$$

25. $(3x^2+8x-4)-(x^2-7x+2) = 3x^2+8x-4-x^2+7x-2$
$$= (3-1)x^2+(8+7)x-4-2$$
$$= 2x^2+15x-6$$

29. $(-7x^3+x^2+6x-12)-(-4x^3-x^2+6x-1) = -7x^3+x^2+6x-12+4x^3+x^2-6x+1$
$$= (-7+4)x^3+(1+1)x^2+(6-6)x-12+1$$
$$= -3x^3+2x^2-11$$

33. $\begin{array}{l}-3a+9\\ \underline{-5a-6}\end{array}$ Add the opposite. $\longrightarrow$ $\begin{array}{l}-3a+9\\ \underline{5a+6}\\ 2a+15\end{array}$

37. $\begin{array}{l}4x^3+6x^2+7x-14\\ \underline{-2x^3-6x^2+7x-9}\end{array}$ Add the opposite. $\longrightarrow$ $\begin{array}{l}4x^3+6x^2+7x-14\\ \underline{2x^3+6x^2-7x+9}\\ 6x^3+12x^2\quad\quad-5\end{array}$

41. $(5x+3)-(7x-2)+(3x+6) = 5x+3-7x+2+3x+6$
$$= (5-7+3)x+3+2+6$$
$$= x+11$$

45. $(x^2-7x-4)+(2x^2-8x-9)-(4x^2-2x-1) = x^2-7x-4+2x^2-8x-9-4x^2+2x+1$
$$= (1+2-4)x^2+(-7-8+2)x-4-9+1$$
$$= -x^2-13x-12$$

49. $(3a-2b)-(7a+4b)-(6a-3b) = 3a-2b-7a-4b-6a+3b$
$$= (3-7-6)a+(-2-4+3)b$$
$$= -10a-3b$$

53. $7x+[3x-(2x-1)] = 7x+[3x-2x+1]$
$$= 7x+[x+1]$$
$$= 7x+x+1$$
$$= 8x+1$$

57. $(5a-1)-[3a+(4a-7)] = 5a-1-[3a+4a-7]$
$$= 5a-1-[7a-7]$$
$$= 5a-1-7a+7$$
$$= -2a+6$$

61. $(4x-2)+(7x+6)-(5x-3) = 4x-2+7x+6-5x+3$
$$= (4+7-5)x-2+6+3$$
$$= 6x+7$$

Problem Set 5.2

1. $(5x)(9x) = 45x^{1+1} = 45x^2$ 5. $(-3xy)(2xy) = -6x^{1+1}y^{1+1} = -6x^2y^2$

9. $(4a^2b^2)(-12ab) = -48a^{2+1}b^{2+1} = -48a^3b^3$

13. $(8ab^2c)(13a^2c) = 104a^{1+2}b^2c^{1+1} = 104a^3b^2c^2$

17. $(4xy)(-2x)(7y^2) = -56x^{1+1}y^{1+2} = -56x^2y^3$

21. $(6cd)(-3c^2d)(-4d) = 72c^{1+2}d^{1+1+1} = 72c^3d^3$

25. $(-\frac{7}{12}a^2b)(\frac{8}{21}b^4) = -\frac{7\cdot\overset{2}{\cancel{8}}}{\underset{3}{\cancel{12}}\cdot\underset{3}{\cancel{21}}}a^2b^{1+4} = -\frac{2}{9}a^2b^5$

29. $(-4ab)(1.6a^3b) = -6.4a^{1+3}b^{1+1} = -6.4a^4b^2$

33. $(-3a^2b^3)^2 = (-3)^2(a^2)^2(b^3)^2 = 9a^4b^6$ 37. $(-4x^4)^3 = (-4)^3(x^4)^3 = -64x^{12}$

41. $(2x^2y)^4 = (2)^4(x^2)^4(y)^4 = 16x^8y^4$

45. $(-x^2y)^6 = (-1)^6(x^2)^6(y)^6 = 1x^{12}y^6 = x^{12}y^6$

49. $3x^2(6x-2) = 3x^2(6x)-3x^2(2) = 18x^3-6x^2$

53. $2x(x^2-4x+6) = 2x(x^2)-2x(4x)+2x(6) = 2x^3-8x^2+12x$

57. $7xy(4x^2-x+5) = 7xy(4x^2)-7xy(x)+7xy(5) = 28x^3y-7x^2y+35xy$

61. $5(x+2y)+4(2x+3y) = 5x+10y+8x+12y = 13x+22y$

65. $2x(x^2-3x-4)+x(2x^2+3x-6) = 2x^3-6x^2-8x+2x^3+3x^2-6x$
$$= 4x^3-3x^2-14x$$

69. $-4(3x+2)-5[2x-(3x+4)] = -4(3x+2)-5[2x-3x-4]$
$$= -4(3x+2)-5(-x-4)$$
$$= -12x-8+5x+20$$
$$= -7x+12$$

73. $(-3x)^3(-4x)^2 = (-27x^3)(16x^2) = -432x^5$

77. $(-a^2bc^3)^3(a^3b)^2 = (-a^6b^3c^9)(a^6b^2)$
$$= -a^{12}b^5c^9$$

Problem Set 5.3

1. $(x+2)(y+3) = x(y+3)+2(y+3) = xy+3x+2y+6$

5. $(x-5)(y-6) = x(y-6)-5(y-6) = xy-6x-5y+30$

9. $(2x+3)(3y+1) = 2x(3y+1)+3(3y+1) = 6xy+2x+9y+3$

13. $(x+8)(x-3) = x(x-3)+8(x-3) = x^2-3x+8x-24$
$$= x^2+5x-24$$

17. $(n-4)(n-6) = n(n-6)-4(n-6) = n^2-6n-4n+24$
$$= n^2-10n+24$$

21. $(5x-2)(3x+7) = 5x(3x+7)-2(3x+7) = 15x^2+35x-6x-14$
$$= 15x^2+29x-14$$

25. $(x+4)(x^2-x-6) = x(x^2-x-6)+4(x^2-x-6)$
$$= x^3-x^2-6x+4x^2-4x-24$$
$$= x^3+3x^2-10x-24$$

29. $(2a-1)(4a^2-5a+9) = 2a(4a^2-5a+9)-1(4a^2-5a+9)$
$$= 8a^3-10a^2+18a-4a^2+5a-9$$
$$= 8a^3-14a^2+23a-9$$

33. $(x^2+2x+3)(x^2+5x+4) = x^2(x^2+5x+4)+2x(x^2+5x+4)+3(x^2+5x+4)$
$$= x^4+5x^3+4x^2+2x^3+10x^2+8x+3x^2+15x+12$$
$$= x^4+7x^3+17x^2+23x+12$$

> For Problems 37-80 you are to use the "shortcut" pattern described in the text for multiplying binomials.

81. Use the pattern $(a+b)^2 = a^2+2ab+b^2$.
$$(x+7)^2 = x^2+2(7)(x)+7^2 = x^2+14x+49$$

85. Use the pattern $(a-b)^2 = a^2-2ab+b^2$.
$$(x-1)^2 = x^2-2(x)(1)+1^2 = x^2-2x+1$$

89. Use the pattern $(a-b)^2 = a^2-2ab+b^2$.
$$(2x-3)^2 = (2x)^2-2(2x)(3)+3^2 = 4x^2-12x+9$$

93. Use the pattern $(a-b)^2 = a^2-2ab+b^2$.
$$(1-5n)^2 = 1^2-2(1)(5n)+(5n)^2 = 1-10n+25n^2$$

97. Use the pattern $(a+b)^2 = a^2+2ab+b^2$.
$$(3+4y)^2 = 3^2+2(3)(4y)+(4y)^2 = 9+24y+16y^2$$

101. Use the pattern $(a-b)^2 = a^2-2ab+b^2$.
$$(4a-7b)^2 = (4a)^2-2(4a)(7b)+(7b)^2 = 16a^2-56ab+49b^2$$

36

105. Use the pattern $(a+b)(a-b) = a^2-b^2$.

$(5x-11y)(5x+11y) = (5x)^2-(11y)^2 = 25x^2-121y^2$

109. $-2x(4x+y)(4x-y) = -2x(16x^2-y^2) = -32x^3+2xy^2$

113. $(x-3)^3 = (x-3)(x-3)(x-3) = (x-3)(x^2-6x+9)$
$$= x^3-6x^2+9x-3x^2+18x-27$$
$$= x^3-9x^2+27x-27$$

117. $(3n-2)^3 = (3n-2)(3n-2)(3n-2)$
$$= (3n-2)(9n^2-12n+4)$$
$$= 27n^3-36n^2+12n-18n^2+24n-8$$
$$= 27n^3-54n^2+36n-8$$

Problem Set 5.4

1. $\dfrac{x^{10}}{x^2} = x^{10-2} = x^8$

5. $\dfrac{-16n^6}{2n^2} = -8n^{6-2} = -8n^4$

9. $\dfrac{65x^2y^3}{5xy} = 13x^{2-1}y^{3-1} = 13xy^2$

13. $\dfrac{18x^2y^6}{xy^2} = 18x^{2-1}y^{6-2} = 18xy^4$

17. $\dfrac{-96x^5y^7}{12y^3} = -8x^5y^{7-3} = -8x^5y^4$

21. $\dfrac{56a^2b^3c^5}{4abc} = 14a^{2-1}b^{3-1}c^{5-1} = 14ab^2c^4$

25. $\dfrac{8x^4+12x^5}{2x^2} = \dfrac{8x^4}{2x^2} + \dfrac{12x^5}{2x^2} = 4x^2+6x^3$

29. $\dfrac{-28n^5+36n^2}{4n^2} = \dfrac{-28n^5}{4n^2} + \dfrac{36n^2}{4n^2} = -7n^3+9$

33. $\dfrac{-24n^8+48n^5-78n^3}{-6n^3} = \dfrac{-24n^8}{-6n^3} + \dfrac{48n^5}{-6n^3} - \dfrac{78n^3}{-6n^3}$
$$= 4n^5-8n^2+13$$

37. $\dfrac{27x^2y^4-45xy^4}{-9xy^3} = \dfrac{27x^2y^4}{-9xy^3} - \dfrac{45xy^4}{-9xy^3}$
$$= -3xy+5y$$

41. $\dfrac{12a^2b^2c^2-52a^2b^3c^5}{-4a^2bc} = \dfrac{12a^2b^2c^2}{-4a^2bc} - \dfrac{52a^2b^3c^5}{-4a^2bc}$
$$= -3bc+13b^2c^4$$

45. $\dfrac{-42x^6-70x^4+98x^2}{14x^2} = \dfrac{-42x^6}{14x^2} - \dfrac{70x^4}{14x^2} + \dfrac{98x^2}{14x^2}$
$$= -3x^4-5x^2+7$$

49. $\dfrac{-xy+5x^2y^3-7x^2y^6}{xy} = \dfrac{-xy}{xy} + \dfrac{5x^2y^3}{xy} - \dfrac{7x^2y^6}{xy}$
$$= -1+5xy^2-7xy^5$$

Problem Set 5.5

1.
$$\begin{array}{r}
x + 12 \\
x+4 \overline{\smash{\big)}\, x^2+16x+48} \\
\underline{x^2+ 4x} \\
12x+48 \\
\underline{12x+48}
\end{array}$$

5.
$$\begin{array}{r}
x + 8 \\
x+3 \overline{\smash{\big)}\, x^2+11x+28} \\
\underline{x^2+ 3x} \\
8x+28 \\
\underline{8x+24} \\
4 \leftarrow \text{Remainder}
\end{array}$$

9.
$$\begin{array}{r}
5n + 4 \\
n-1 \overline{\smash{\big)}\, 5n^2 - n-4} \\
\underline{5n^2-5n} \\
4n-4 \\
\underline{4n-4}
\end{array}$$

13.
$$\begin{array}{r}
4x - 7 \\
5x+1 \overline{\smash{\big)}\, 20x^2-31x-7} \\
\underline{20x^2 + 4x} \\
-35x-7 \\
\underline{-35x-7}
\end{array}$$

17.
$$\begin{array}{r}
2x^2 + 3x + 4 \\
x-2 \overline{\smash{\big)}\, 2x^3 - x^2-2x-8} \\
\underline{2x^3-4x^2} \\
3x^2-2x-8 \\
\underline{3x^2-6x} \\
4x-8 \\
\underline{4x-8}
\end{array}$$

21.
$$\begin{array}{r}
n^2 + 6n - 4 \\
n-6 \overline{\smash{\big)}\, n^3+0n^2-40n+24} \\
\underline{n^3-6n^2} \\
6n^2-40n+24 \\
\underline{6n^2-36n} \\
-4n+24 \\
\underline{-4n+24}
\end{array}$$

25.
$$\begin{array}{r}
9x^2 + 12x + 16 \\
3x-4 \overline{\smash{\big)}\, 27x^3 + 0x^2 + 0x-64} \\
\underline{27x^3-36x^2} \\
36x^2 + 0x-64 \\
\underline{36x^2-48x} \\
48x-64 \\
\underline{48x-64}
\end{array}$$

29.
$$\begin{array}{r}
3t + 2 \\
3t-1 \overline{\smash{\big)}\, 9t^2+3t+4} \\
\underline{9t^2-3t} \\
6t+4 \\
\underline{6t-2} \\
6 \leftarrow \text{Remainder}
\end{array}$$

33.
$$\begin{array}{r}
4x^2 - 5x + 5 \\
x+7 \overline{\smash{\big)}\, 4x^3+23x^2-30x+32} \\
\underline{4x^3+28x^2} \\
-5x^2-30x+32 \\
\underline{-5x^2-35x} \\
5x+32 \\
\underline{5x+35} \\
-3 \leftarrow \text{Remainder}
\end{array}$$

37.
$$\begin{array}{r}
2x - 12 \\
x^2+4x \overline{\smash{\big)}\, 2x^3 - 4x^2 + x-5} \\
\underline{2x^3 + 8x^2} \\
-12x^2 + x-5 \\
\underline{-12x^2-48x} \\
49x-5 \leftarrow \text{Remainder}
\end{array}$$

Problem Set 5.6

1. $3^{-2} = \dfrac{1}{3^2} = \dfrac{1}{9}$

5. $\left(\dfrac{3}{2}\right)^{-1} = \dfrac{1}{\left(\dfrac{3}{2}\right)^1} = \dfrac{1}{\frac{3}{2}} = \dfrac{2}{3}$

9. $\left(-\dfrac{4}{3}\right)^0 = 1$ Any nonzero number raised to the zero power is one.

13. $(-2)^{-2} = \dfrac{1}{(-2)^2} = \dfrac{1}{4}$

17. $\dfrac{1}{\left(\frac{3}{4}\right)^{-3}} = \dfrac{1}{\frac{1}{\left(\frac{3}{4}\right)^3}} = \left(\dfrac{3}{4}\right)^3 = \dfrac{27}{64}$

21. $3^6 \cdot 3^{-3} = 3^{6+(-3)} = 3^3 = 27$

25. $\dfrac{10^{-1}}{10^2} = 10^{-1-2} = 10^{-3} = \dfrac{1}{10^3} = \dfrac{1}{1000}$

29. $\left(\dfrac{4^{-1}}{3}\right)^{-2} = \dfrac{4^2}{3^{-2}} = 4^2 \cdot 3^2 = 144$

33. $n^{-4}n^2 = n^{-4+2} = n^{-2} = \dfrac{1}{n^2}$

37. $(2x^3)(4x^{-2}) = 8x^{3+(-2)} = 8x$

41. $(5y^{-1})(-3y^{-2}) = -15y^{-1+(-2)} = -15y^{-3} = -\dfrac{15}{y^3}$

45. $\dfrac{x^7}{x^{-3}} = x^{7-(-3)} = x^{7+3} = x^{10}$

49. $\dfrac{4n^{-1}}{2n^{-3}} = 2n^{-1-(-3)} = 2n^{1+3} = 2n^2$

53. $\dfrac{-52y^{-2}}{-13y^{-2}} = 4y^{-2-(-2)} = 4y^{-2+2} - 4y^0 = 4(1) = 4$

57. $(x^2)^{-2} = x^{-2(2)} = x^{-4} = \dfrac{1}{x^4}$

61. $(x^{-2}y^{-1})^3 = (x^{-2})^3 (y^{-1})^3 = x^{-6}y^{-3}$
$$= \dfrac{1}{x^6 y^3}$$

65. $(4n^3)^{-2} = (4)^{-2}(n^3)^{-2} = (4)^{-2}n^{-6} = \dfrac{1}{(4^2)n^6} = \dfrac{1}{16n^6}$

69. $(5x^{-1})^{-2} = (5)^{-2}(x^{-1})^{-2} = (5)^{-2}x^2 = \dfrac{x^2}{5^2} = \dfrac{x^2}{25}$

73. $\left(\dfrac{x^2}{y}\right)^{-1} = \dfrac{x^{-2}}{y^{-1}} = \dfrac{y}{x^2}$

77. $\left(\dfrac{x^{-1}}{y^{-3}}\right)^{-2} = \dfrac{x^2}{y^6}$

81. $\left(\dfrac{2x^{-1}}{x^{-2}}\right)^{-3} = (2x)^{-3} = \dfrac{1}{(2x)^3} = \dfrac{1}{8x^3}$

For Problems 85–106, you need to refer to the text for the rules for changing back and forth between standard decimal notation and scientific notation.

109. $(5,000,000)(.00009) = (5)(10^6)(9)(10^{-5})$
$$= (45)(10)$$
$$= 450$$

113. $\dfrac{.00086}{4300} = \dfrac{(8.6)(10^{-4})}{(4.3)(10^{3})} = 2(10^{-7}) = .0000002$

117. $\dfrac{(.0008)(.07)}{(20000)(.0004)} = \dfrac{(8)(10^{-4})(7)(10^{-2})}{(2)(10^{4})(4)(10^{-4})} = \dfrac{(56)(10^{-6})}{(8)(10^{0})}$

$= (7)(10^{-6})$

$= .000007$

Chapter 6

Problem Set 6.1

1. $24y = 2 \cdot 2 \cdot 2 \cdot 3 \cdot y$
$30xy = 2 \cdot 3 \cdot 5 \cdot x \cdot y$
The greatest common factor is $2 \cdot 3 \cdot y = 6y$.

5. $42ab^3 = 2 \cdot 3 \cdot 7 \cdot a \cdot b \cdot b \cdot b$
$70a^2b^2 = 2 \cdot 5 \cdot 7 \cdot a \cdot a \cdot b \cdot b$
The greatest common factor is $2 \cdot 7 \cdot a \cdot b \cdot b = 14ab^2$.

9. $16a^2b^2 = 2 \cdot 2 \cdot 2 \cdot 2 \cdot a \cdot a \cdot b \cdot b$
$40a^2b^3 = 2 \cdot 2 \cdot 2 \cdot 5 \cdot a \cdot a \cdot b \cdot b \cdot b$
$56a^3b^4 = 2 \cdot 2 \cdot 2 \cdot 7 \cdot a \cdot a \cdot a \cdot b \cdot b \cdot b \cdot b$
The greatest common factor is $2 \cdot 2 \cdot 2 \cdot a \cdot a \cdot b \cdot b = 8a^2b^2$.

13. $14xy-21y = 7y(2x)-7y(3) = 7y(2x-3)$

17. $12xy^2-30x^2y = 6xy(2y)-6xy(5x) = 6xy(2y-5x)$

21. $16xy^3+25x^2y^2 = xy^2(16y)+xy^2(25x) = xy^2(16y+25x)$

25. $9a^2b^4-27a^2b = 9a^2b(b^3)-9a^2b(3) = 9a^2b(b^3-3)$

29. $40x^2y^2+8x^2y = 8x^2y(5y)+8x^2y(1) = 8x^2y(5y+1)$

33. $2x^3-3x^2+4x = x(2x^2)-x(3x)+x(4) = x(2x^2-3x+4)$

37. $14a^2b^3+35ab^2-49a^3b = 7ab(2ab^2)+7ab(5b)-7ab(7a^2)$
$= 7ab(2ab^2+5b-7a^2)$

41. $a(b-4)-c(b-4) = (b-4)(a-c)$

45. $2x(x+1)-3(x+1) = (x+1)(2x-3)$

49. $bx-by-cx+cy = b(x-y)-c(x-y)$
$= (x-y)(b-c)$

53. $x^2+5x+12x+60 = x(x+5)+12(x+5)$
$= (x+5)(x+12)$

57. $2x^2+x-10x-5 = x(2x+1)-5(2x+1)$
$= (2x+1)(x-5)$

61. $x^2-8x = 0$
$x(x-8) = 0$
$x = 0$ or $x-8 = 0$
$x = 0$ or $x = 8$

The solution set is $\{0,8\}$.

65. $n^2 = 5n$
$n^2-5n = 0$
$n(n-5) = 0$
$n = 0$ or $n-5 = 0$
$n = 0$ or $n = 5$

The solution set is $\{0,5\}$.

69. $7x^2 = -3x$
$7x^2+3x = 0$
$x(7x+3) = 0$
$x = 0$ or $7x+3 = 0$
$x = 0$ or $7x = -3$
$x = 0$ or $x = -\frac{3}{7}$

The solution set is $\{-\frac{3}{7},0\}$.

73. $\quad 4x^2 = 6x$

$\qquad 2x^2 = 3x \qquad$ Divide both sides by 2.

$\quad 2x^2-3x = 0$

$\quad x(2x-3) = 0$

$\quad x = 0$ or $2x-3 = 0$

$\quad x = 0$ or $\quad 2x = 3$

$\quad x = 0$ or $\quad x = \dfrac{3}{2}$

The solution set is $\{0,\dfrac{3}{2}\}$.

77. $\qquad 13x = x^2$

$\quad 13x-x^2 = 0$

$\quad x(13-x) = 0$

$\quad x = 0$ or $13-x = 0$

$\quad x = 0$ or $\quad 13 = x$

The solution set is $\{0,13\}$.

81. Let n represent the number.

$\qquad n^2 = 9n$

$\quad n^2-9n = 0$

$\quad n(n-9) = 0$

$\quad n = 0$ or $n-9 = 0$

$\quad n = 0$ or $\quad n = 9$

The number is 0 or 9.

85. Let s represent the length of a side of the square and therefore also represents the length of a radius of the circle.

$\qquad \Pi s^2 = 4s$

$\quad \Pi s^2-4s = 0$

$\quad s(\Pi s-4) = 0$

$\quad s = 0$ or $\Pi s-4 = 0$

$\quad s = 0$ or $\quad \Pi s = 4$

$\quad s = 0$ or $\quad s = \dfrac{4}{\Pi}$

The answer of 0 must be discarded, so the length of a side of the square and the length of a radius of the circle is $\dfrac{4}{\Pi}$ units.

Problem Set 6.2

1. $x^2-1 = x^2-1^2 = (x-1)(x+1)$

5. $x^2-4y^2 = x^2-(2y)^2 = (x-2y)(x+2y)$

9. $36a^2-25b^2 = (6a)^2-(5b)^2 = (6a-5b)(6a+5b)$

13. $5x^2-20 = 5(x^2-4) = 5(x-2)(x+2)$

17. $2x^2-18y^2 = 2(x^2-9y^2) = 2(x-3y)(x+3y)$

21. x^2+9y^2 is not factorable.

25. $36-4x^2 = 4(9-x^2) = 4(3-x)(3+x)$

29. $x^4-81 = (x^2-9)(x^2+9) = (x-3)(x+3)(x^2+9)$

33. $3x^3+48x = 3x(x^2+16)$

37. $4x^2-64 = 4(x^2-16) = 4(x-4)(x+4)$

41. $\qquad x^2 = 9$

$\quad x^2-9 = 0$

$\quad (x-3)(x+3) = 0$

$\quad x-3 = 0$ or $x+3 = 0$

$\quad x = 3$ or $\quad x = -3$

The solution set is $\{-3,3\}$.

45. $\qquad 9x^2 = 16$

$\quad 9x^2-16 = 0$

$\quad (3x-4)(3x+4) = 0$

$\quad 3x-4 = 0$ or $3x+4 = 0$

$\quad 3x = 4$ or $\quad 3x = -4$

$\quad x = \dfrac{4}{3}$ or $\quad x = -\dfrac{4}{3}$

The solution set is $\{-\dfrac{4}{3},\dfrac{4}{3}\}$.

49.
$$25x^2 = 4$$
$$25x^2-4 = 0$$
$$(5x-2)(5x+2) = 0$$
$$5x-2 = 0 \text{ or } 5x+2 = 0$$
$$5x = 2 \text{ or } 5x = -2$$
$$x = \frac{2}{5} \text{ or } x = -\frac{2}{5}$$

The solution set is $\{-\frac{2}{5}, \frac{2}{5}\}$.

53.
$$3x^3-48x = 0$$
$$x^3-16x = 0 \qquad \text{Divide both sides by 3.}$$
$$x(x^2-16) = 0$$
$$x(x-4)(x+4) = 0$$
$$x = 0 \text{ or } x-4 = 0 \text{ or } x+4 = 0$$
$$x = 0 \text{ or } x = 4 \text{ or } x = -4$$

The solution set is $\{-4,0,4\}$.

57.
$$5-45x^2 = 0$$
$$1-9x^2 = 0 \qquad \text{Divide both sides by 5.}$$
$$(1-3x)(1+3x) = 0$$
$$1-3x = 0 \text{ or } 1+3x = 0$$
$$-3x = -1 \text{ or } 3x = -1$$
$$x = \frac{1}{3} \text{ or } x = -\frac{1}{3}$$

The solution set is $\{-\frac{1}{3}, \frac{1}{3}\}$.

61.
$$64x^2 = 81$$
$$64x^2-81 = 0$$
$$(8x-9)(8x+9) = 0$$
$$8x-9 = 0 \text{ or } 8x+9 = 0$$
$$8x = 9 \text{ or } 8x = -9$$
$$x = \frac{9}{8} \text{ or } x = -\frac{9}{8}$$

The solution set is $\{-\frac{9}{8}, \frac{9}{8}\}$.

65. Let n represent the number.

$$n^2-49 = 0$$
$$(n-7)(n+7) = 0$$
$$n-7 = 0 \text{ or } n+7 = 0$$
$$n = 7 \text{ or } n = -7$$

The number is -7 or 7.

69. Let s represent the length of a side of the smaller square. Then 5s represents the length of a side of the larger square.

$$s^2+(5s)^2 = 234$$
$$s^2+25s^2 = 234$$
$$26s^2 = 234$$
$$s^2 = 9$$
$$s^2-9 = 0$$
$$(s-3)(s+3) = 0$$
$$s-3 = 0 \text{ or } s+3 = 0$$
$$s = 3 \text{ or } s = -3$$

The negative solution must be discarded since we are working with the lengths of line segments. Thus, the small square is 3 inches by 3 inches and the large square is 15 inches by 15 inches.

73. Let r and 2r represent the length of a radius of each circle.

$$\Pi r^2 + \Pi(2r)^2 = 80\Pi$$
$$r^2 + 4r^2 = 80$$
$$5r^2 = 80$$
$$r^2 = 16$$
$$r^2 - 16 = 0$$
$$(r-4)(r+4) = 0$$
$$r-4 = 0 \text{ or } r+4 = 0$$
$$r = 4 \text{ or } \quad r = -4$$

Again we must discard the negative solution. The radii are of length 4 meters and 8 meters.

Problem Set 6.3

1. $x^2 + 10 + 24$ We need two integers whose sum is 10 and whose product is 24. They are 4 and 6.

$$x^2 + 10x + 24 = (x+4)(x+6)$$

5. $x^2 - 11x + 18$ We need two integers whose sum is -11 and whose product is 18. They are -2 and -9.

$$x^2 - 11x + 18 = (x-2)(x-9)$$

9. $n^2 + 6n - 27$ We need two integers whose sum is 6 and whose product is -27. They are 9 and -3.

$$n^2 + 6n - 27 = (n+9)(n-3)$$

13. $t^2 + 12t + 24$ We need two integers whose sum is 12 and whose product is 24. The possible factors of 24 are 1(24), 2(12), 3(8), and 4(6). None of these pairs have a sum of 12. Therefore, $t^2 + 12t + 24$ is not factorable using integers.

17. $x^2 + 5x - 66$ We need two integers whose sum is 5 and whose product is -66. They are 11 and -6.

$$x^2 + 5x - 66 = (x+11)(x-6)$$

21. $x^2 + 21x + 80$ We need two integers whose sum is 21 and whose product is 80. They are 16 and 5.

$$x^2 + 21x + 80 = (x+16)(x+5)$$

25. $x^2 - 10x - 48$ We need two integers whose sum is -10 and whose product is -48. The possible factors of -48 are -1(48), 1(-48), -2(24), 2(-24), -3(16), 3(-16), -4(12), 4(-12), -6(8), and 6(-8). None of these have a sum of -10. Therefore, $x^2 - 10x - 48$ is not factorable.

29. $a^2 - 4ab - 32b^2$ We need two integers whose sum is -4 and whose product is -32. They are -8 and 4.

$$a^2 - 4ab - 32b^2 = (a-8b)(a+4b)$$

33.　　　　$x^2-9x+18 = 0$
　　　　　$(x-3)(x-6) = 0$
　　　　　$x-3 = 0$ or $x-6 = 0$
　　　　　　　$x = 3$ or　$x = 6$

　　　The solution set is $\{3,6\}$.

41.　　　　$t^2+t-56 = 0$
　　　　　$(t+8)(t-7) = 0$
　　　　　$t+8 = 0$ or $t-7 = 0$
　　　　　　$t = -8$ or　$t = 7$

　　　The solution set is $\{-8,7\}$.

37.　　　　$n^2+5n-36 = 0$
　　　　　$(n+9)(n-4) = 0$
　　　　　$n+9 = 0$ or $n-4 = 0$
　　　　　　$n = -9$ or　$n = 4$

　　　The solution set is $\{-9,4\}$.

45.　　　　　$x^2+11x = 12$
　　　　　$x^2+11x-12 = 0$
　　　　　$(x+12)(x-1) = 0$
　　　　　$x+12 = 0$ or $x-1 = 0$
　　　　　　$x = -12$ or　$x = 1$

　　　The solution set is $\{-12,1\}$.

49.　　　　$-x^2-2x+24 = 0$
　　　　　$x^2+2x-24 = 0$　　Multiply both sides by -1.
　　　　　$(x+6)(x-4) = 0$
　　　　　$x+6 = 0$ or $x-4 = 0$
　　　　　　$x = -6$ or　$x = 4$

　　　The solution set is $\{-6,4\}$.

53.　Let x and $x+2$ represent the consecutive even whole numbers.

　　　　　$x(x+2) = 168$
　　　　$x^2+2x-168 = 0$
　　　$(x+14)(x-12) = 0$
　　　$x+14 = 0$ or $x-12 = 0$
　　　　$x = -14$ or　$x = 12$

　　　The negative solution must be discarded since we are looking for whole numbers. Thus, the numbers are 12 and $12+2 = 14$.

57.　Let n and $n-3$ represent the numbers.

　　　　　$n^2 = 10(n-3)+9$
　　　　　$n^2 = 10n-30+9$
　　　　　$n^2 = 10n-21$
　　　$n^2-10n+21 = 0$
　　　$(n-7)(n-3) = 0$
　　　$n-7 = 0$ or $n-3 = 0$
　　　　$n = 7$ or　$n = 3$

　　　The numbers are 7 and 4 or 3 and 0.

61.　Let w represent the width and $15-w$ the length.

　　　　　$w(15-w) = 54$
　　　　　$15w-w^2 = 54$
　　　　　　$0 = w^2-15w+54$
　　　　　　$0 = (w-9)(w-6)$
　　　　$w-9 = 0$ or $w-6 = 0$
　　　　　$w = 9$ or　$w = 6$

　　　The rectangle is 9 centimeters by 6 centimeters.

65. Let x represent the length of the longer leg. Then x−7 represents the length of the shorter leg and x+2 represents the length of the hypotenuse.

$$x^2+(x-7)^2 = (x+2)^2$$
$$x^2+x^2-14x+49 = x^2+4x+4$$
$$x^2-18x+45 = 0$$
$$(x-15)(x-3) = 0$$
$$x-15 = 0 \text{ or } x-3 = 0$$
$$x = 15 \text{ or } x = 3$$

The solution of 3 must be discarded since that would yield a negative answer for x−7. Thus, one leg is 15 feet long, the other leg is 15−7 = 8 feet long, and the hypotenuse is 15+2 = 17 feet long.

Problem Set 6.4

In the text we demonstrated two techniques for factoring trinomials of the form ax^2+bx+c. The first technique relied heavily on trial and error along with your knowledge of multiplying binomials. The second technique was a bit more systematic. If you are able to find the factors using the first technique, that is great. However, if you run into trouble, then try the more systematic second technique. We will use that second technique here to help you out.

1. $3x^2+7x+2$ sum of 7

product of 3(2) = 6

We need two integers whose sum is 7 and whose product is 6. They are 1 and 6. Now the middle term, 7x, can be written as x+6x and we can factor as follows.

$$3x^2+7x+2 = 3x^2+x+6x+2$$
$$= x(3x+1)+2(3x+1)$$
$$= (3x+1)(x+2)$$

5. $4x^2-25x+6$ sum of −25

product of 4(6) = 24

We need two integers whose sum is −25 and whose product is 24. They are −1 and −24.

$$4x^2-25x+6 = 4x^2-x-24x+6$$
$$= x(4x-1)-6(4x-1)$$
$$= (4x-1)(x-6)$$

46

9. $5y^2-33y-14$ sum of -33

product of $5(-14) = -70$

We need two integers whose sum is -33 and whose product is -70. They are -35 and 2.

$$5y^2-33y-14 = 5y^2-35y+2y-14$$
$$= 5y(y-7)+2(y-7)$$
$$= (y-7)(5y+2)$$

13. $2x^2+x+7$ sum of 1

product of $2(7) = 14$

We need two integers whose sum is 1 and whose product is 14. It should be evident that no two integers can satisfy these conditions. Therefore, $2x^2+x+7$ is not factorable.

17. $7x^2-30x+8$ sum of -30

product of $7(8) = 56$

We need two integers whose sum is -30 and whose product is 56. They are -28 and -2.

$$7x^2-20x+8 = 7x^2-28x-2x+8$$
$$= 7x(x-4)-2(x-4)$$
$$= (x-4)(7x-2)$$

21. $9t^2-15t-14$ sum of -15

product of $9(-14) = -126$

We need two integers whose sum is -15 and whose product is -126. They are -21 and 6.

$$9t^2-15t-14 = 9t^2-21t+6t-14$$
$$= 3t(3t-7)+2(3t-7)$$
$$= (3t-7)(3t+2)$$

25. $6n^2+2n-5$ sum of 2

product of $6(-5) = -30$

We need two integers whose sum is 2 and whose product is -30. The possible pairs of factors of -30 are $1(-30)$, $-1(30)$, $2(-15)$, $-2(15)$, $3(-10)$, $-3(10)$, $6(-5)$, and $5(-6)$. None of these pairs of factors have a sum of 2. Therefore, $6n^2+2n-5$ is not factorable.

29. $20x^2-31x+12$ sum of -31

product of $20(12) = 240$

We need two integers whose sum is -31 and whose product is 240. They are -16 and -15.

$$20x^2-31x+12 = 20x^2-16x-15x+12$$
$$= 4x(5x-4)-3(5x-4)$$
$$= (5x-4)(4x-3)$$

33. $24x^2-50x+25$ sum of -50

product of $24(25) = 600$

We need two integers whose sum is -50 and whose product is 600. They are -30 and -20.

$$24x^2-50x+25 = 24x^2-30x-20x+25$$
$$= 6x(4x-5)-5(4x-5)$$
$$= (4x-5)(6x-5)$$

37. $21a^2+a-2$ sum of 1

product of $21(-2) = -42$

We need two integers whose sum is 1 and whose product is -42. They are 7 and -6.

$$21a^2+a-2 = 21a^2+7a-6a-2$$
$$= 7a(3a+1)-2(3a+1)$$
$$= (3a+1)(7a-2)$$

41. $4x^2+12x+9$ sum of 12

product of $4(9) = 36$

We need two integers whose sum is 12 and whose product is 36. They are 6 and 6.

$$4x^2+12x+9 = 4x^2+6x+6x+9$$
$$= 2x(2x+3)+3(2x+3)$$
$$= (2x+3)(2x+3)$$

45. $20x^2+7xy-6y^2$ — sum of 7

$$\text{product of } 20(-6) = -120$$

We need two integers whose sum is 7 and whose product is -120. They are 15 and -8.

$$\begin{aligned}
20x^2+7xy-6y^2 &= 20x^2+15xy-8xy-6y^2 \\
&= 5x(4x+3y)-2y(4x+3y) \\
&= (4x+3y)(5x-2y)
\end{aligned}$$

49. $8x^2-55x-7$ — sum of -55

$$\text{product of } 8(-7) = -56$$

We need two integers whose sum is -55 and whose product is -56. They are -56 and 1.

$$\begin{aligned}
8x^2-55x-7 &= 8x^2-56x+x-7 \\
&= 8x(x-7)+1(x-7) \\
&= (x-7)(8x+1)
\end{aligned}$$

53.
$$12x^2+11x+2 = 0$$
$$(4x+1)(3x+2) = 0$$
$$4x+1 = 0 \text{ or } 3x+2 = 0$$
$$4x = -1 \text{ or } \quad 3x = -2$$
$$x = -\frac{1}{4} \text{ or } \quad x = -\frac{2}{3}$$

The solution set is $\{-\frac{2}{3},-\frac{1}{4}\}$.

57.
$$15n^2-41n+14 = 0$$
$$(3n-7)(5n-2) = 0$$
$$3n-7 = 0 \text{ or } 5n-2 = 0$$
$$3n = 7 \text{ or } \quad 5n = 2$$
$$n = \frac{7}{3} \text{ or } \quad n = \frac{2}{5}$$

The solution set is $\{\frac{2}{5},\frac{7}{3}\}$.

61.
$$16y^2-18y-9 = 0$$
$$(8y+3)(2y-3) = 0$$
$$8y+3 = 0 \quad \text{or } 2y-3 = 0$$
$$8y = -3 \text{ or } \quad 2y = 3$$
$$y = -\frac{3}{8} \text{ or } \quad y = \frac{3}{2}$$

The solution set is $\{-\frac{3}{8},\frac{3}{2}\}$.

65.
$$10x^2-29x+10 = 0$$
$$(2x-5)(5x-2) = 0$$
$$2x-5 = 0 \text{ or } 5x-2 = 0$$
$$2x = 5 \text{ or } \quad 5x = 2$$
$$x = \frac{5}{2} \text{ or } \quad x = \frac{2}{5}$$

The solution set is $\{\frac{2}{5},\frac{5}{2}\}$.

69.
$$16x(x+1) = 5$$
$$16x^2+16x-5 = 0$$
$$(4x-1)(4x+5) = 0$$
$$4x-1 = 0 \text{ or } 4x+5 = 0$$
$$4x = 1 \text{ or } \quad 4x = -5$$
$$x = \frac{1}{4} \text{ or } x = -\frac{5}{4} \ .$$

The solution set is $\{-\frac{5}{4},\frac{1}{4}\}$.

73.
$$4x^2-45x+50 = 0$$
$$(4x-5)(x-10) = 0$$
$$4x-5 = 0 \text{ or } x-10 = 0$$
$$4x = 5 \text{ or } \quad x = 10$$
$$x = \frac{5}{4} \text{ or } \quad x = 10$$

The solution set is $\{\frac{5}{4},10\}$.

77.
$$12x^2-43x-20 = 0$$
$$(12x+5)(x-4) = 0$$
$$12x+5 = 0 \text{ or } x-4 = 0$$
$$12x = -5 \text{ or } \quad x = 4$$
$$x = -\frac{5}{12} \text{ or } \quad x = 4$$

The solution set is $\{-\frac{5}{12}, 4\}$.

Problem Set 6.5

1. $x^2+4x+4 = (x+2)(x+2) = (x+2)^2$ **5.** $9n^2+12n+4 = (3n+2)(3n+2) = (3n+2)^2$

9. $4+36x+81x^2 = (2+9x)(2+9x) = (2+9x)^2$

13. $2x^2+17x+8$ → sum of 17

product of $2(8) = 16$

We need two integers whose sum is 17 and whose product is 16. They are 16 and 1.

$$2x^2+17x+8 = 2x^2+16x+x+8$$
$$= 2x(x+8)+1(x+8)$$
$$= (x+8)(2x+1)$$

17. $n^2-7n-60$ We need two integers whose sum is -7 and whose product is -60. They are -12 and 5.

$$n^2-7n-60 = (n-12)(n+5)$$

21. $8x^2+72 = 8(x^2)+8(9) = 8(x^2+9)$ **25.** $15x^2+65x+70 = 5(3x^2+13x+14)$
$$= 5(3x+7)(x+2)$$

29. $xy+5y-8x-40 = y(x+5)-8(x+5)$ **33.** $24x^2+18x-81 = 3(8x^2+6x-27)$
$$= (x+5)(y-8) \qquad\qquad\qquad\qquad = 3(4x+9)(2x-3)$$

37. $5x^4-80 = 5(x^4-16) = 5(x^2-4)(x^2+4)$
$$= 5(x-2)(x+2)(x^2+4)$$

41.
$$4x^2-20x = 0$$
$$4x(x-5) = 0$$
$$4x = 0 \text{ or } x-5 = 0$$
$$x = 0 \text{ or } \quad x = 5$$

The solution set is $\{0,5\}$.

45.
$$-2x^3+8x = 0$$
$$-2x(x^2-4) = 0$$
$$-2x(x-2)(x+2) = 0$$
$$-2x = 0 \text{ or } x-2 = 0 \text{ or } x+2 = 0$$
$$x = 0 \text{ or } \quad x = 2 \text{ or } \quad x = -2$$

The solution set is $\{-2,0,2\}$.

49.
$$(3n-1)(4n-3) = 0$$
$$3n-1 = 0 \text{ or } 4n-3 = 0$$
$$3n = 1 \text{ or } \quad 4n = 3$$
$$n = \frac{1}{3} \text{ or } \quad n = \frac{3}{4}$$

The solution set is $\{\frac{1}{3}, \frac{3}{4}\}$.

53.
$$2x^2 = 12x$$
$$x^2 = 6x \qquad \text{Divide both sides by 2.}$$
$$x^2-6x = 0$$
$$x(x-6) = 0$$
$$x = 0 \text{ or } x-6 = 0$$
$$x = 0 \text{ or } \quad x = 6$$

The solution set is $\{0,6\}$.

50

57.
$$12-40x+25x^2 = 0$$
$$(2-5x)(6-5x) = 0$$
$$2-5x = 0 \quad \text{or} \quad 6-5x = 0$$
$$-5x = -2 \quad \text{or} \quad -5x = -6$$
$$x = \frac{2}{5} \quad \text{or} \quad x = \frac{6}{5}$$

The solution set is $\{\frac{2}{5}, \frac{6}{5}\}$.

61.
$$(3n+1)(n+2) = 12$$
$$3n^2+7n+2 = 12$$
$$3n^2+7n-10 = 0$$
$$(3n+10)(n-1) = 0$$
$$3n+10 = 0 \quad \text{or} \quad n-1 = 0$$
$$3n = -10 \quad \text{or} \quad n = 1$$
$$n = -\frac{10}{3} \quad \text{or} \quad n = 1$$

The solution set is $\{-\frac{10}{3}, 1\}$.

65.
$$9x^2-24x+16 = 0$$
$$(3x-4)^2 = 0$$
$$3x-4 = 0$$
$$3x = 4$$
$$x = \frac{4}{3}$$

The solution set is $\{\frac{4}{3}\}$.

69.
$$24x^2+17x-20 = 0$$
$$(8x-5)(3x+4) = 0$$
$$8x-5 = 0 \quad \text{or} \quad 3x+4 = 0$$
$$8x = 5 \quad \text{or} \quad 3x = -4$$
$$x = \frac{5}{8} \quad \text{or} \quad x = -\frac{4}{3}$$

The solution set is $\{-\frac{4}{3}, \frac{5}{8}\}$.

73. Let n and 2n+3 represent the numbers.
$$n(2n+3) = -1$$
$$2n^2+3n+1 = 0$$
$$(2n+1)(n+1) = 0$$
$$2n+1 = 0 \quad \text{or} \quad n+1 = 0$$
$$2n = -1 \quad \text{or} \quad n = -1$$
$$n = -\frac{1}{2} \quad \text{or} \quad n = -1$$

If $n = -\frac{1}{2}$ then $2n+3 = 2(-\frac{1}{2})+3 = 2$.

If $n = 1$ then $2n+3 = 2(-1)+3 = 1$.

77. Let n represent the number of rows. Then 2n-3 represents the number of chairs per row.
$$n(2n-3) = 54$$
$$2n^2-3n-54 = 0$$
$$(2n+9)(n-6) = 0$$
$$2n+9 = 0 \quad \text{or} \quad n-6 = 0$$
$$2n = -9 \quad \text{or} \quad n = 6$$
$$n = -\frac{9}{2} \quad \text{or} \quad n = 6$$

The negative solution must be discarded since n represents the number of rows of chairs. Thus, there are 6 rows and 2(6)-3 = 9 chairs per row.

81. Let w represent the width of the rectangle. Then 2w+1 represents its length.
$$w(2w+1) = 55$$
$$2w^2+w-55 = 0$$
$$(2w+11)(w-5) = 0$$
$$2w+11 = 0 \quad \text{or} \quad w-5 = 0$$
$$2w = -11 \quad \text{or} \quad w = 5$$
$$w = -\frac{11}{2} \quad \text{or} \quad w = 5$$

The negative solution must be discarded. Therefore, the rectangle is 5 centimeters wide and 2(5)+1 = 11 centimeters long.

85. Let x represent the width of the strip.

$$(8-2x)(11-2x) = 40$$
$$88-38x+4x^2 = 40$$
$$4x^2-38x+48 = 0$$
$$2x^2-19x+24 = 0$$
$$(2x-3)(x-8) = 0$$
$$2x-3 = 0 \text{ or } x-8 = 0$$
$$2x = 3 \text{ or } \quad x = 8$$
$$x = \frac{3}{2} \text{ or } \quad x = 8$$

The solution of 8 must be discarded since it would not be possible to cut a strip 8 inches wide from both sides of a piece that is only 8 inches wide to start with. Therefore, the strip to be cut off is $1\frac{1}{2}$ inches wide.

Chapter 7

Problem Set 7.1

1. $\dfrac{6x}{14y} = \dfrac{2 \cdot 3x}{2 \cdot 7y} = \dfrac{3x}{7y}$

5. $\dfrac{-15x^2 y}{25x} = -\dfrac{5x \cdot 3xy}{5x \cdot 5} = -\dfrac{3xy}{5}$

9. $\dfrac{12a^2 b^5}{-54a^2 b^3} = -\dfrac{\cancel{6} \cdot 2 \cdot \cancel{a^2} \cdot b^{\cancel{5}}^{\,b^2}}{\cancel{6} \cdot 9 \cdot \cancel{a^2} \cdot \cancel{b^3}} = -\dfrac{2b^2}{9}$

13. $\dfrac{xy}{x^2 - 2x} = \dfrac{\cancel{x}y}{\cancel{x}(x-2)} = \dfrac{y}{x-2}$

17. $\dfrac{x^2 + 2x}{x^2 - 7x} = \dfrac{\cancel{x}(x+2)}{\cancel{x}(x-7)} = \dfrac{x+2}{x-7}$

21. $\dfrac{15 - 3n}{n-5} = \dfrac{3(5-n)}{n-5} = 3(-1) = -3$

25. $\dfrac{x^2 - 1}{3x^2 - 3x} = \dfrac{(\cancel{x-1})(x+1)}{3x(\cancel{x-1})} = \dfrac{x+1}{3x}$

29. $\dfrac{6x^3 - 15x^2 y}{6x^2 + 24xy} = \dfrac{\cancel{3x^2}^{\,x}(2x - 5y)}{\cancel{6x}_{2}(x + 4y)} = \dfrac{x(2x - 5y)}{2(x + 4y)}$

33. $\dfrac{2n^2 + 5n - 3}{n^2 - 9} = \dfrac{(2n-1)\cancel{(n+3)}}{(n-3)\cancel{(n+3)}} = \dfrac{2n-1}{n-3}$

37. $\dfrac{9(x-1)^2}{12(x-1)^3} = \dfrac{\cancel{9}^{3}\cancel{(x-1)}\cancel{(x-1)}}{\cancel{12}_{4}\cancel{(x-1)}\cancel{(x-1)}(x-1)} = \dfrac{3}{4(x-1)}$

41. $\dfrac{10a^2 + a - 3}{15a^2 + 4a - 3} = \dfrac{(5a+3)(2a-1)}{(5a+3)(3a-1)} = \dfrac{2a-1}{3a-1}$

45. $\dfrac{x^2 - 9}{-x^2 - 3x} = \dfrac{(x+3)\cancel{(x-3)}}{-x\cancel{(x-3)}} = \dfrac{x+3}{-x} = -\dfrac{x+3}{x}$

49. $\dfrac{4n^2 - 12n + 9}{2n^2 - n - 3} = \dfrac{\cancel{(2n-3)}(2n-3)}{\cancel{(2n-3)}(n+1)} = \dfrac{2n-3}{n+1}$

53. $\dfrac{1 - x^2}{x - x^2} = \dfrac{(1-x)(1+x)}{x(1-x)} = \dfrac{1+x}{x}$

57. $\dfrac{x^2 + 7x - 18}{12 - 4x - x^2} = \dfrac{(x+9)(x-2)}{(6+x)(2-x)} = -\dfrac{x+9}{x+6}$ (Don't forget that $\dfrac{x-2}{2-x} = -1$.)

Problem Set 7.2

1. $\dfrac{5}{9} \cdot \dfrac{3}{10} = \dfrac{\cancel{5}^{\,1} \cdot \cancel{3}^{\,1}}{\cancel{9}_{3} \cdot \cancel{10}_{2}} = \dfrac{1}{6}$

5. $\left(\dfrac{17}{9}\right) \div \left(-\dfrac{19}{9}\right) = \left(\dfrac{17}{9}\right)\left(-\dfrac{9}{19}\right) = -\dfrac{17}{19}$

9. $\left(-\dfrac{5n^2}{18n}\right)\left(\dfrac{27n}{25}\right) = -\dfrac{\cancel{5} \cdot \cancel{27}^{3} \cdot n^{\cancel{3}^{\,n^2}}}{\cancel{18} \cdot \cancel{25}_{5} \cdot \cancel{n}} = -\dfrac{3n^2}{10}$

13. $\dfrac{18a^2 b^2}{-27a} \div \dfrac{-9a}{5b} = \left(-\dfrac{18a^2 b^2}{27a}\right)\left(-\dfrac{5b}{9a}\right)$

$= \dfrac{18 \cdot 5 \cdot \cancel{a^2}^{\,2} \cdot b^3}{27 \cdot \cancel{9} \cdot \cancel{a^2}} = \dfrac{10b^3}{27}$

17. $\dfrac{1}{15ab^3} \div \dfrac{-1}{12a} = \left(\dfrac{1}{15ab^3}\right)\left(-\dfrac{12a}{1}\right) = -\dfrac{\cancel{12a}^{4}}{\cancel{15ab^3}_{5}} = -\dfrac{4}{5b^3}$

21. $\dfrac{y}{x+y} \cdot \dfrac{x^2 - y^2}{xy} = \dfrac{\cancel{y}(x-y)\cancel{(x+y)}}{x\cancel{y}\cancel{(x+y)}} = \dfrac{x-y}{x}$

25. $\dfrac{6ab}{4ab+4b^2} \div \dfrac{7a-7b}{a^2-b^2} = \dfrac{\overset{3}{\cancel{6}ab}}{\underset{2}{\cancel{4b}(a+b)}} \cdot \dfrac{(a+b)(a-b)}{7(a-b)} = \dfrac{3a}{14}$

29. $\dfrac{2x^2-3xy+y^2}{4x^2y} \div \dfrac{x^2-y^2}{6x^2y^2} = \dfrac{(2x-y)(x-y)}{\underset{2}{\cancel{4x^2y}}} \cdot \dfrac{\overset{3}{\cancel{6x^2}}\overset{y}{\cancel{y^2}}}{(x+y)(x-y)} = \dfrac{3y(2x-y)}{2(x+y)}$

33. $\dfrac{2x^2-2xy}{x^2+4x-32} \cdot \dfrac{x^2-16}{5xy-5y^2} = \dfrac{2x(x-y)(x-4)(x+4)}{(x+8)(x-4)5y(x-y)} = \dfrac{2x(x+4)}{5y(x+8)}$

37. $\dfrac{(3t-1)^2}{45t-15} \div \dfrac{12t^2+5t-3}{20t+5} = \dfrac{(3t-1)(3t-1)}{\underset{3}{\cancel{15}(3t-1)}} \cdot \dfrac{\overset{5}{\cancel{5}}(4t+1)}{(3t-1)(4t+3)} = \dfrac{4t+1}{3(4t+3)}$

41. $\dfrac{6}{9y} \div \dfrac{30x}{12y^2} \cdot \dfrac{5xy}{4} = \dfrac{6}{9y} \cdot \dfrac{12y^2}{30x} \cdot \dfrac{5xy}{4}$

$$= \dfrac{6 \cdot \overset{2}{\cancel{12}} \cdot \cancel{3} \cdot \cancel{x} \cdot \overset{y^2}{\cancel{y}}}{9 \cdot \underset{3}{\cancel{30}} \cdot \underset{\cancel{6}}{4} \cdot \cancel{x} \cdot \cancel{y}} = \dfrac{y^2}{3}$$

45. $\dfrac{x^2+9x+18}{x^2+3x} \cdot \dfrac{x^2+5x}{x^2-25} \div \dfrac{x^2+8x}{x^2+3x-40} = \dfrac{(x+6)(x+3)}{x(x+3)} \cdot \dfrac{x(x+5)}{(x+5)(x-5)} \cdot \dfrac{(x+8)(x-5)}{x(x+8)} = \dfrac{x+6}{x}$

Problem Set 7.3

1. $\dfrac{5}{x} + \dfrac{12}{x} = \dfrac{5+12}{x} = \dfrac{17}{x}$

5. $\dfrac{7}{2n} + \dfrac{1}{2n} = \dfrac{7+1}{2n} = \dfrac{8}{2n} = \dfrac{4}{n}$

9. $\dfrac{x+1}{x} + \dfrac{3}{x} = \dfrac{x+1+3}{x} = \dfrac{x+4}{x}$

13. $\dfrac{x+1}{x} - \dfrac{1}{x} = \dfrac{x+1-1}{x} = \dfrac{x}{x} = 1$

17. $\dfrac{7a+2}{3} - \dfrac{4a-6}{3} = \dfrac{7a+2-(4a-6)}{3} = \dfrac{7a+2-4a+6}{3} = \dfrac{3a+8}{3}$

21. $\dfrac{3n-7}{6} - \dfrac{9n-1}{6} = \dfrac{3n-7-(9n-1)}{6} = \dfrac{3n-7-9n+1}{6}$

$$= \dfrac{-6n-6}{6} = \dfrac{-6(n+1)}{6} = -(n+1) = -n-1$$

25. $\dfrac{3(x+2)}{4x} + \dfrac{6(x-1)}{4x} = \dfrac{3x+6+6x-6}{4x} = \dfrac{9x}{4x} = \dfrac{9}{4}$

29. $\dfrac{2(3x-4)}{7x^2} - \dfrac{7x-8}{7x^2} = \dfrac{6x-8-7x+8}{7x^2} = \dfrac{-x}{7x^2} = -\dfrac{1}{7x}$

33. $\dfrac{3x}{(x-6)^2} - \dfrac{18}{(x-6)^2} = \dfrac{3x-18}{(x-6)^2} = \dfrac{3(x-6)}{(x-6)^2} = \dfrac{3}{x-6}$

37. $\dfrac{7n}{12} - \dfrac{4n}{3} = \dfrac{7n}{12} - \left(\dfrac{4}{4}\right)\left(\dfrac{4n}{3}\right) = \dfrac{7n}{12} - \dfrac{16n}{12} = -\dfrac{9n}{12} = -\dfrac{3n}{4}$

41. $\dfrac{8x}{3} - \dfrac{3x}{7} = \left(\dfrac{7}{7}\right)\left(\dfrac{8x}{3}\right) - \left(\dfrac{3}{3}\right)\left(\dfrac{3x}{7}\right) = \dfrac{56x}{21} - \dfrac{9x}{21} = \dfrac{56x-9x}{21} = \dfrac{47x}{21}$

45. $\frac{7n}{8} - \frac{3n}{9} = \frac{7n}{8} - \frac{n}{3} = (\frac{3}{3})(\frac{7n}{8}) - (\frac{8}{8})(\frac{n}{3})$

$$= \frac{21n}{24} - \frac{8n}{24} = \frac{21n-8n}{24} = \frac{13n}{24}$$

49. $\frac{x-6}{9} + \frac{x+2}{3} = \frac{x-6}{9} + (\frac{3}{3})(\frac{x+2}{3}) = \frac{x-6}{9} + \frac{3x+6}{9}$

$$= \frac{x-6+3x+6}{9} = \frac{4x}{9}$$

53. $\frac{4n-3}{6} - \frac{3n+5}{18} = (\frac{3}{3})(\frac{4n-3}{6}) - \frac{3n+5}{18}$

$$= \frac{12n-9}{18} - \frac{3n+5}{18} = \frac{12n-9-3n-5}{18} = \frac{9n-14}{18}$$

57. $\frac{x}{5} - \frac{3}{10} - \frac{7x}{12} = (\frac{12}{12})(\frac{x}{5}) - (\frac{6}{6})(\frac{3}{10}) - (\frac{5}{5})(\frac{7x}{12})$

$$= \frac{12x}{60} - \frac{18}{60} - \frac{35x}{60}$$

$$= \frac{12x-18-35x}{60} = \frac{-23x-18}{60}$$

61. $\frac{5}{6y} - \frac{7}{9y} = (\frac{3}{3})(\frac{5}{6y}) - (\frac{2}{2})(\frac{7}{9y}) = \frac{15}{18y} - \frac{14}{18y} = \frac{15-14}{18y} = \frac{1}{18y}$

65. $\frac{3}{2x} - \frac{2}{3x} + \frac{5}{4x} = (\frac{6}{6})(\frac{3}{2x}) - (\frac{4}{4})(\frac{2}{3x}) + (\frac{3}{3})(\frac{5}{4x}) = \frac{18}{12x} - \frac{8}{12x} + \frac{15}{12x}$

$$= \frac{18-8+15}{12x} = \frac{25}{12x}$$

69. $\frac{2}{n-1} - \frac{3}{n} = (\frac{n}{n})(\frac{2}{n-1}) - (\frac{n-1}{n-1})(\frac{3}{n}) = \frac{2n}{n(n-1)} - \frac{3(n-1)}{n(n-1)} = \frac{2n-3n+3}{n(n-1)} = \frac{-n+3}{n(n-1)}$

73. $\frac{6}{x} - \frac{12}{2x+1} = (\frac{2x+1}{2x+1})(\frac{6}{x}) - (\frac{x}{x})(\frac{12}{2x+1}) = \frac{6(2x+1)-12x}{x(2x+1)}$

$$= \frac{12x+6-12x}{x(2x+1)} = \frac{6}{x(2x+1)}$$

77. $\frac{3}{x-2} - \frac{9}{x+1} = (\frac{x+1}{x+1})(\frac{3}{x-2}) - (\frac{x-2}{x-2})(\frac{9}{x+1}) = \frac{3(x+1)-9(x-2)}{(x+1)(x-2)}$

$$= \frac{3x+3-9x+18}{(x+1)(x-2)} = \frac{-6x+21}{(x+1)(x-2)}$$

Problem Set 7.4

1. Use $x(x-4)$ as the LCD.

$$\frac{4}{x(x-4)} + \frac{3}{x} = \frac{4}{x(x-4)} + (\frac{x-4}{x-4})(\frac{3}{x}) = \frac{4+3x-12}{x(x-4)} = \frac{3x-8}{x(x-4)}$$

5. Use $n(n-6)$ as the LCD.

$$\frac{8}{n} - \frac{2}{n(n-6)} = (\frac{n-6}{n-6})(\frac{8}{n}) - \frac{2}{n(n-6)} = \frac{8n-48-2}{n(n-6)} = \frac{8n-50}{n(n-6)}$$

9. Use $2x(x-1)$ as the LCD.

$$\frac{7}{2x} - \frac{x}{x(x-1)} = (\frac{x-1}{x-1})(\frac{7}{2x}) - (\frac{2}{2})\left(\frac{x}{x(x-1)}\right) = \frac{7x-7-2x}{2x(x-1)} = \frac{5x-7}{2x(x-1)}$$

55

13. Use $(x+1)(x-1)$ as the LCD.

$$\frac{8x}{(x-1)(x+1)} - \frac{4}{x-1} = \frac{8x}{(x-1)(x+1)} - (\frac{x+1}{x+1})(\frac{4}{x-1}) = \frac{8x-4x-4}{(x-1)(x+1)} = \frac{4x-4}{(x-1)(x+1)}$$

$$= \frac{4(x-1)}{(x-1)(x+1)} = \frac{4}{x+1}$$

17. $\left.\begin{array}{l} x^2-6x = x(x-6) \\ x^2+6x = x(x+6) \end{array}\right\}$ Use $x(x-6)(x+6)$ as the LCD.

$$\frac{1}{x(x-6)} - \frac{1}{x(x+6)} = (\frac{x+6}{x+6})\left(\frac{1}{x(x-6)}\right) - (\frac{x-6}{x-6})\left(\frac{1}{x(x+6)}\right)$$

$$= \frac{x+6-x+6}{x(x+6)(x-6)} = \frac{12}{x(x+6)(x-6)}$$

21. $\left.\begin{array}{l} 6x+4 = 2(3x+2) \\ 9x+6 = 3(3x+2) \end{array}\right\}$ Use $6(3x+2)$ as the LCD.

$$\frac{5x}{2(3x+2)} + \frac{2x}{3(3x+2)} = (\frac{3}{3})\left(\frac{5x}{2(3x+2)}\right) + (\frac{2}{2})\left(\frac{2x}{3(3x+2)}\right) = \frac{15x+4x}{6(3x+2)} = \frac{19x}{6(3x+2)}$$

25. $\left.\begin{array}{l} x^2+7x+12 = (x+3)(x+4) \\ x^2-9 = (x-3)(x+3) \end{array}\right\}$ Use $(x+3)(x-3)(x+4)$ as the LCD.

$$\frac{2}{(x+3)(x+4)} + \frac{3}{(x-3)(x+3)} = (\frac{x-3}{x-3})\left(\frac{2}{(x+3)(x+4)}\right) + (\frac{x+4}{x+4})\left(\frac{3}{(x-3)(x+3)}\right)$$

$$= \frac{2(x-3)+3(x+4)}{(x-3)(x+3)(x+4)} = \frac{2x-6+3x+12}{(x-3)(x+3)(x+4)} = \frac{5x+6}{(x-3)(x+3)(x+4)}$$

29. $\left.\begin{array}{l} ab+b^2 = b(a+b) \\ a^2+ab = a(a+b) \end{array}\right\}$ Use $ab(a+b)$ as the LCD.

$$\frac{a}{b(a+b)} - \frac{b}{a(a+b)} = (\frac{a}{a})\left(\frac{a}{b(a+b)}\right) - (\frac{b}{b})\left(\frac{b}{a(a+b)}\right) = \frac{a^2-b^2}{ab(a+b)} = \frac{(a+b)(a-b)}{ab(a+b)} = \frac{a-b}{ab}$$

33. $\left.\begin{array}{l} x^2-2x = x(x-2) \\ x^2+2x = x(x+2) \\ x^2-4 = (x-2)(x+2) \end{array}\right\}$ Use $x(x-2)(x+2)$ as the LCD.

$$\frac{10}{x(x-2)} + \frac{8}{x(x+2)} - \frac{3}{(x+2)(x-2)} = (\frac{x+2}{x+2})\left(\frac{10}{x(x-2)}\right) + (\frac{x-2}{x-2})\left(\frac{8}{x(x+2)}\right) - (\frac{x}{x})\left(\frac{3}{(x-2)(x+2)}\right)$$

$$= \frac{10(x+2)+8(x-2)-3x}{x(x-2)(x+2)}$$

$$= \frac{10x+20+8x-16-3x}{x(x-2)(x+2)} = \frac{15x+4}{x(x-2)(x+2)}$$

37. Use $(3x-5)(x+4)$ as the LCD.

$$\frac{5x}{(3x-5)(x+4)} - \frac{1}{3x-5} - \frac{2}{x+4}$$

$$= \frac{5x}{(3x-5)(x+4)} - \frac{(x+4)}{(x+4)}\left(\frac{1}{3x-5}\right) - \left(\frac{3x-5}{3x-5}\right)\left(\frac{2}{x+4}\right)$$

$$= \frac{5x-(x+4)-2(3x-5)}{(3x-5)(x+4)} = \frac{5x-x-4-6x+10}{(3x-5)(x+4)} = \frac{-2x+6}{(3x-5)(x+4)}$$

⎡Be sure that you understand both techniques demonstrated in Example 7⎤
⎣in the text before you begin working Problems 41-60.   ⎦

41. $\left(\dfrac{\frac{1}{2} - \frac{3}{4}}{\frac{1}{6} + \frac{1}{3}}\right)\left(\dfrac{12}{12}\right) = \dfrac{12\left(\frac{1}{2} - \frac{3}{4}\right)}{12\left(\frac{1}{6} + \frac{1}{3}\right)} = \dfrac{6-9}{2+4} = -\dfrac{3}{6} = -\dfrac{1}{2}$

45. $\left(\dfrac{3 - \frac{2}{3}}{2 + \frac{1}{4}}\right)\left(\dfrac{12}{12}\right) = \dfrac{12\left(3 - \frac{2}{3}\right)}{12\left(2 + \frac{1}{4}\right)} = \dfrac{36-8}{24+3} = \dfrac{28}{27}$

49. $\left(\dfrac{\frac{2}{x} + \frac{3}{y}}{\frac{5}{x} - \frac{1}{y}}\right)\left(\dfrac{xy}{xy}\right) = \dfrac{xy\left(\frac{2}{x} + \frac{3}{y}\right)}{xy\left(\frac{5}{x} - \frac{1}{y}\right)} = \dfrac{2y+3x}{5y-x}$

53. $\left(\dfrac{\frac{6}{x} + 2}{\frac{3}{x} + 4}\right)\left(\dfrac{x}{x}\right) = \dfrac{x\left(\frac{6}{x} + 2\right)}{x\left(\frac{3}{x} + 4\right)} = \dfrac{6+2x}{3+4x}$

57. $\left(\dfrac{\frac{x+2}{4}}{\frac{1}{x} + \frac{3}{2}}\right)\left(\dfrac{4x}{4x}\right) = \dfrac{4x\left(\frac{x+2}{4}\right)}{4x\left(\frac{1}{x} + \frac{3}{2}\right)} = \dfrac{x^2+2x}{4+6x}$

Problem Set 7.5

1. $6\left(\dfrac{x}{2} + \dfrac{x}{3}\right) = 6(10)$ Multiply both sides by 6.

       $3x+2x = 60$

           $5x = 60$

            $x = 12$

The solution set is $\{12\}$.

5. $6\left(\dfrac{n}{2} + \dfrac{n-1}{6}\right) = 6\left(\dfrac{5}{2}\right)$ Multiply both sides by 6.

       $3n+n-1 = 15$

         $4n-1 = 15$

           $4n = 16$

            $n = 4$

The solution set is $\{4\}$.

9. $12\left(\dfrac{2x+3}{3} + \dfrac{3x-4}{4}\right) = 12\left(\dfrac{17}{4}\right)$ Multiply both sides by 12.

     $4(2x+3)+3(3x-4) = 51$

       $8x+12+9x-12 = 51$

              $17x = 51$

               $x = 3$

The solution set is $\{3\}$.

13.
$$30\left(\frac{3x+2}{5} - \frac{2x-1}{6}\right) = 30\left(\frac{2}{15}\right) \quad \text{Multiply both sides by 30.}$$
$$6(3x+2)-5(2x-1) = 4$$
$$18x+12-10x+5 = 4$$
$$8x+17 = 4$$
$$8x = -13$$
$$x = -\frac{13}{8}$$

The solution set is $\{-\frac{13}{8}\}$.

17.
$$\frac{5}{3n} - \frac{1}{9} = \frac{1}{n}, \quad n \neq 0$$
$$9n\left(\frac{5}{3n} - \frac{1}{9}\right) = 9n\left(\frac{1}{n}\right) \quad \text{Multiply both sides by 9n.}$$
$$15-n = 9$$
$$-n = -6$$
$$n = 6$$

The solution set is $\{6\}$.

21.
$$\frac{4}{5t} - 1 = \frac{3}{2t}, \quad t \neq 0$$
$$10t\left(\frac{4}{5t} - 1\right) = 10t\left(\frac{3}{2t}\right) \quad \text{Multiply both sides by 10t.}$$
$$8-10t = 15$$
$$-10t = 7$$
$$t = -\frac{7}{10}$$

The solution set is $\{-\frac{7}{10}\}$.

25.
$$\frac{90-n}{n} = 10 + \frac{2}{n}, \quad n \neq 0$$
$$n\left(\frac{90-n}{n}\right) = n\left(10 + \frac{2}{n}\right) \quad \text{Multiply both sides by n.}$$
$$90-n = 10n+2$$
$$88 = 11n$$
$$8 = n$$

The solution set is $\{8\}$.

29.
$$\frac{x}{x+3} - 2 = \frac{-3}{x+3}, \quad x \neq -3$$
$$(x+3)\left(\frac{x}{x+3} - 2\right) = (x+3)\left(\frac{-3}{x+3}\right)$$
$$x-2(x+3) = -3$$
$$x-2x-6 = -3$$
$$-x = 3$$
$$x = -3$$
Our initial restriction was $x \neq -3$.
Therefore, the solution set is $\emptyset$.

33.
$$\frac{x}{x+2} + 3 = \frac{1}{x+2}, \quad x \neq -2$$
$$(x+2)\left(\frac{x}{x+2} + 3\right) = (x+2)\left(\frac{1}{x+2}\right)$$
$$x+3(x+2) = 1$$
$$x+3x+6 = 1$$
$$4x = 5$$
$$x = -\frac{5}{4}$$

The solution set is $\{-\frac{5}{4}\}$.

37.
$$1 + \frac{n+1}{2n} = \frac{3}{4}, \quad n \neq 0$$
$$4n\left(1 + \frac{n+1}{2n}\right) = 4n\left(\frac{3}{4}\right) \quad \text{Multiply both sides by 4n.}$$
$$4n+2(n+1) = 3n$$
$$4n+2n+2 = 3n$$
$$6n+2 = 3n$$
$$3n = -2$$
$$n = -\frac{2}{3}$$

The solution set is $\{-\frac{2}{3}\}$.

41. Let x represent the denominator and x-8 the numerator.
$$\frac{x-8}{x} = \frac{5}{6}$$
$$6(x-8) = 5x$$
$$6x-48 = 5x$$
$$x = 48$$

The fraction is $\frac{40}{48}$.

45. Let x represent the smaller number and 65-x the larger number.
$$\frac{65-x}{x} = 8 + \frac{2}{x}$$
$$65-x = 8x+2$$
$$63 = 9x$$
$$7 = x$$

The numbers are 7 and 65-7 = 58.

49.

	rate	time	distance
Heidi	x	$\dfrac{125}{x}$	125
Abby	x	$\dfrac{75}{x}$	75

Since Heidi's time is $3\frac{1}{3}$ hours more than Abby's time, the following equation can be set up and solved.

$$\frac{125}{x} = \frac{75}{x} + \frac{10}{3}$$
$$375 = 225+10x$$
$$150 = 10x$$
$$15 = x$$

Their rates were 15 miles per hour.

Problem Set 7.6

1. $\dfrac{4}{x} + \dfrac{7}{6} = \dfrac{1}{x} + \dfrac{2}{3x}$, $x \neq 0$

$6x(\dfrac{4}{x} + \dfrac{7}{6}) = 6x(\dfrac{1}{x} + \dfrac{2}{3x})$ Multiply both
sides by 6x.

$$24+7x = 6+4$$
$$24+7x = 10$$
$$7x = -14$$
$$x = -2$$

The solution set is $\{-2\}$.

5. $\dfrac{5}{2(n-5)} - \dfrac{3}{n-5} = 1$, $n \neq 5$

$2(n-5)[\dfrac{5}{2(n-5)} - \dfrac{3}{n-5}] = 2(n-5)(1)$ Multiply both sides by $2(n-5)$.

$$5-6 = 2n-10$$
$$-1 = 2n-10$$
$$9 = 2n$$
$$\frac{9}{2} = n$$

The solution set is $\{\frac{9}{2}\}$.

9. $\dfrac{x}{x-2} + \dfrac{4}{x+2} = 1$, $x \neq 2$ and $x \neq -2$

$(x-2)(x+2)[\dfrac{x}{x-2} + \dfrac{4}{x+2}] = (x-2)(x+2)(1)$ Multiply both sides by $(x-2)(x+2)$.

$$x(x+2)+4(x-2) = x^2-4$$
$$x^2+2x+4x-8 = x^2-4$$
$$6x-8 = -4$$
$$6x = 4$$
$$x = \frac{4}{6} = \frac{2}{3}$$

The solution set is $\{\frac{2}{3}\}$.

13.
$$\frac{3n}{n+3} - \frac{n}{n-3} = 2 \ , \quad n \neq -3 \text{ and } n \neq 3$$

$$(n+3)(n-3)[\frac{3n}{n+3} - \frac{n}{n-3}] = (n+3)(n-3)(2) \quad \text{Multiply both sides by } (n+3)(n-3).$$
$$3n(n-3)-n(n+3) = 2(n^2-9)$$
$$3n^2-9n-n^2-3n = 2n^2-18$$
$$-12n = -18$$
$$n = \frac{-18}{-12} = \frac{3}{2}$$

The solution set is $\{\frac{3}{2}\}$.

17.
$$\frac{4}{x-1} - \frac{2x-3}{x^2-1} = \frac{6}{x+1} \ , \quad x \neq 1 \text{ and } x \neq -1$$

$$(x-1)(x+1)[\frac{4}{x-1} - \frac{2x-3}{x^2-1}] = (x-1)(x+1)[\frac{6}{x+1}] \quad \text{Multiply both sides by } (x-1)(x+1).$$
$$4(x+1)-(2x-3) = 6(x-1)$$
$$4x+4-2x+3 = 6x-6$$
$$2x+7 = 6x-6$$
$$13 = 4x$$
$$\frac{13}{4} = x$$

The solution set is $\{\frac{13}{4}\}$.

21.
$$n + \frac{1}{n} = \frac{17}{4} \ , \quad n \neq 0$$

$$4n(n + \frac{1}{n}) = 4n(\frac{17}{4}) \quad \text{Multiply both sides by } 4n.$$
$$4n^2+4 = 17n$$
$$4n^2-17n+4 = 0$$
$$(4n-1)(n-4) = 0$$
$$4n-1 = 0 \text{ or } n-4 = 0$$
$$4n = 1 \text{ or } \quad n = 4$$
$$n = \frac{1}{4} \text{ or } \quad n = 4$$

The solution set is $\{\frac{1}{4}, 4\}$.

25.
$$x - \frac{5x}{x-2} = \frac{-10}{x-2} \ , \quad x \neq 2$$

$$(x-2)[x - \frac{5x}{x-2}] = (x-2)[\frac{-10}{x-2}] \quad \text{Multiply both sides by } x-2.$$
$$x(x-2)-5x = -10$$
$$x^2-2x-5x+10 = 0$$
$$x^2-7x+10 = 0$$
$$(x-2)(x-5) = 0$$
$$x-2 = 0 \text{ or } x-5 = 0$$
$$x = 2 \text{ or } \quad x = 5$$

The solution of 2 must be discarded because of the initial restriction, $x \neq 2$. Thus, the solution set is $\{5\}$.

29.
$$\frac{3}{n-5} + \frac{4}{n+7} = \frac{2n+11}{n^2+2n-35} \quad , \quad n \neq 5 \text{ and } n \neq -7$$

$$(n-5)(n+7)\left[\frac{3}{n-5} + \frac{4}{n+7}\right] = (n-5)(n+7)\left[\frac{2n+11}{(n-5)(n+7)}\right]$$

$$3(n+7)+4(n-5) = 2n+11$$
$$3n+21+4n-20 = 2n+11$$
$$7n+1 = 2n+11$$
$$5n = 10$$
$$n = 2$$

The solution set is {2}.

33. Let n represent the number and $\frac{1}{n}$ its reciprocal.

$$n+2\left(\frac{1}{n}\right) = \frac{9}{2}$$
$$2n^2+4 = 9n$$
$$2n^2-9x+4 = 0$$
$$(2n-1)(n-4) = 0$$
$$2n-1 = 0 \text{ or } n-4 = 0$$
$$2n = 1 \text{ or } \quad n = 4$$
$$n = \frac{1}{2} \text{ or } \quad n = 4$$

The numbers are $\frac{1}{2}$ or 4.

37.

	rate	time	distance
Celia	$\frac{60}{x-2}$	x-2	60
Tom	$\frac{85}{x}$	x	85

Since Celia's rate is 3 miles per hour faster than Tom's rate, we can set up and solve the following equation.

$$\frac{60}{x-2} = \frac{85}{x} + 3$$
$$60x = 85(x-2)+3x(x-2)$$
$$60x = 85x-170+3x^2-6x$$
$$0 = 3x^2+19x-170$$
$$0 = (3x+34)(x-5)$$
$$3x+34 = 0 \text{ or } x-5 = 0$$
$$3x = -34 \text{ or } \quad x = 5$$
$$x = -\frac{34}{3} \text{ or } \quad x = 5$$

The negative solution must be discarded. Thus, Tom's time is 5 hours and Celia's time is 3 hours. Therefore, Tom's rate is $\frac{85}{5}$ = 17 miles per hour and Celia's rate is $\frac{60}{3}$ = 20 miles per hour.

41. Let m represent the time that it takes for the tank to overflow.

$$\frac{m}{5} - \frac{m}{6} = 1$$

$$6m - 5m = 30$$
$$m = 30$$

It will take 30 minutes.

45. Let h represent the time that it would take Mike by himself.

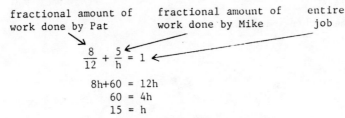

$$\frac{8}{12} + \frac{5}{h} = 1$$

$$8h + 60 = 12h$$
$$60 = 4h$$
$$15 = h$$

It would take Mike 15 hours by himself.

49. Let m represent Paul's time and m−5 represent Amelia's time. Then Paul's rate is $\frac{600}{n}$ and Amelia's rate is $\frac{600}{m-5}$. Since Amelia's rate is 20 words per minute faster than Paul's rate, we can set up and solve the following equation.

$$\frac{600}{m-5} = \frac{600}{m} + 20$$

$$600m = 600(m-5) + 20m(m-5)$$
$$600m = 600m - 3000 + 20m^2 - 100m$$
$$0 = 20m^2 - 100m - 3000$$
$$0 = m^2 - 5m - 150$$
$$0 = (m-15)(m+10)$$
$$m - 15 = 0 \quad \text{or} \quad m+10 = 0$$
$$m = 15 \quad \text{or} \quad m = -10$$

The negative solution must be discarded. Thus, Paul's time is 15 minutes and his rate is 600÷15 = 40 words per minute. Amelia's time is 15−5 = 10 minutes and her rate is 600÷10 = 60 words per minute.

Chapter 8

Problem Set 8.1

1. $3x+7y = 13$
    $7y = 13-3x$
    $y = \dfrac{13-3x}{7}$

5. $-x+5y = 14$
    $5y = x+14$
    $y = \dfrac{x+14}{5}$

9. $-2x+3y = -5$
    $3y = -5+2x$
    $y = \dfrac{-5+2x}{3}$

For Problems 11-34 you need to plot a sufficient number of points to determine the figure.

Problem Set 8.2

For each of these straight line graphs,

   (1) find the y-intercept by letting x = 0 and solve for y,

   (2) find the x-intercept by letting y = 0 and solve for x,

   (3) find an additional "check" point,

   (4) plot the three points and connect them with a straight line.

If the line contains the origin or is parallel to an axis, two points in addition to the one intercept should be plotted.

Problem Set 8.3

1. Let (7,5) be P_1 and (3,2) be P_2.

    $m = \dfrac{2-5}{3-7} = \dfrac{-3}{-4} = \dfrac{3}{4}$

5. Let (2,8) be P_1 and (7,2) be P_2.

    $m = \dfrac{2-8}{7-2} = \dfrac{-6}{5} = -\dfrac{6}{5}$

9. Let (4,-1) be P_1 and (-4,-7) be P_2.

    $m = \dfrac{-7-(-1)}{-4-4} = \dfrac{-6}{-8} = \dfrac{3}{4}$

13. Let (-6,-1) be P_1 and (-2,-7) be P_2.

    $m = \dfrac{-7-(-1)}{-2-(-6)} = \dfrac{-6}{4} = -\dfrac{3}{2}$

17. Let (-1,10) be P_1 and (-9,2) be P_2.

    $m = \dfrac{2-10}{-9-(-1)} = \dfrac{-8}{-8} = 1$

21. $\dfrac{y-8}{2-7} = \dfrac{4}{5}$

    $\dfrac{y-8}{-5} = \dfrac{4}{5}$
    $5(y-8) = 4(-5)$
    $5y-40 = -20$
    $5y = 20$
    $y = 4$

41. $3x+2y = 6$ If x = 0, then 2y = 6 and y = 3. If y = 0, then 3x = 6 and x = 2. The points (0,3) and (2,0) can be used to determine the slope.

    $m = \dfrac{0-3}{2-0} = -\dfrac{3}{2}$

45. $x+5y = 6$ If $x = 0$, then $5y = 6$ and $y = \frac{6}{5}$. If $y = 0$, then $x = 6$. The points $(0,\frac{6}{5})$ and $(6,0)$ can be used to determine the slope.

$$m = \frac{0 - \frac{6}{5}}{6} = \frac{-\frac{6}{5}}{6} = -\frac{1}{5}$$

49. $y = 3$ The points $(0,3)$ and $(2,3)$ are on the line and can be used to determine the slope.

$$m = \frac{3-3}{2-0} = \frac{0}{2} = 0$$

53. $6x-5y = 30$ If $x = 0$, then $-5y = -30$ and $y = 6$. If $y = 0$, then $6x = -30$ and $x = -5$. The points $(0,6)$ and $(-5,0)$ can be used to determine the slope.

$$m = \frac{6-0}{0-(-5)} = \frac{6}{5}$$

57. $y = 4x$ If $x = 0$ then $y = 0$. If $x = 1$, then $y = 4$. The points $(0,0)$ and $(1,4)$ can be used to determine the slope of the line.

$$m = \frac{4-0}{1-0} = \frac{4}{1} = 4$$

Problem Set 8.4

⌈Problems 1-12 can be done by using the general approach demonstrated in⌉
⌊Example 1 in the text or by using the point-slope form.⌋

1. Let's use the general approach for this problem. Since the slope determined by $(2,3)$ and (x,y) is $\frac{2}{3}$, we have

$$\frac{y-3}{x-2} = \frac{2}{3}$$
$$2(x-2) = 3(y-3)$$
$$2x-4 = 3y-9$$
$$2x-3y = -5.$$

5. Let's use the point-slope form for this problem.

$$y-y_1 = m(x-x_1)$$
$$y-8 = -\frac{1}{3}(x+4)$$
$$3y-24 = -x-4$$
$$x+3y = 20$$

9. Let's use the point-slope form for this problem.

$$y-y_1 = m(x-x_1)$$
$$y-0 = -\frac{4}{9}(x-0)$$
$$y = -\frac{4}{9}x$$
$$9y = -4x$$
$$4x+9y = 0$$

13. The points $(2,3)$ and $(7,10)$ can be used to determine the slope.

$$m = \frac{10-3}{7-2} = \frac{7}{5}$$

Now we can use either point and the slope in the point-slope form.

$$y-y_1 = m(x-x_1)$$
$$y-3 = \frac{7}{5}(x-2)$$
$$5y-15 = 7x-14$$
$$-1 = 7x-5y$$

17. The points $(-1,-2)$ and $(-6,-7)$ can be used to determine the slope.

$$m = \frac{-2-(-7)}{-1-(-6)} = \frac{5}{5} = 1$$

Now we can use either point and the slope in the point-slope form.

$$y-y_1 = m(x-x_1)$$
$$y+2 = 1(x+1)$$
$$y+2 = x+1$$
$$1 = x-y$$

21. The points $(0,4)$ and $(7,0)$ can be used to determine the slope.

$$m = \frac{4-0}{0-7} = \frac{4}{-7} = -\frac{4}{7}$$

Now we can use either point and the slope in the point-slope form.

$$y-y_1 = m(x-x_1)$$
$$y-4 = -\frac{4}{7}(x-0)$$
$$7y-28 = -4x$$
$$4x+7y = 28$$

25. We can use the slope-intercept form. Substitute 2 for m and -1 for b.

$$y = mx+b$$
$$y = 2x-1$$
$$1 = 2x-y$$

29. Substitute -1 for m and $\frac{5}{2}$ for b in the slope-intercept form.

$$y = mx+b$$
$$y = -x+\frac{5}{2}$$
$$2y = -2x+5$$
$$2x+2y = 5$$

33. $y = -2x-5$
    $\quad\;\;\uparrow\;\;\;\uparrow$
    $\quad\;\;m\;\;\;b$

The slope is -2 and the y-intercept is -5.

37. $-4x+9y = 18$
    $9y = 4x+18$
    $y = \frac{4}{9}x+2$
    $\quad\uparrow\qquad\uparrow$
    $\quad m\qquad b$ The slope is $\frac{4}{9}$ and the y-intercept is 2.

41. $-2x-11y = 11$
$-11y = 2x+11$

$$y = -\frac{2}{11}x - 1$$

$$\underset{m}{\uparrow} \qquad \underset{b}{\uparrow}$$

The slope is $-\frac{2}{11}$ and the y-intercept is -1.

Problem Set 8.5

1. We need to check if $(1,4)$ satisfies both equations.

$5x+y = 9 \longrightarrow 5(1)+4 = 9 \qquad$ Yes

$3x-2y = 4 \longrightarrow 3(1)-2(4) = 4 \quad$ No

Therefore, $(1,4)$ is not a solution of the system.

5. We need to check if $(-1,-2)$ satisfies both equations.

$y = 2x \longrightarrow -2 = 2(-1) \qquad\qquad$ Yes

$3x-4y = 5 \longrightarrow 3(-1)-4(-2) = 5 \quad$ Yes

Therefore, $(-1,-2)$ is a solution of the system.

9. We need to check if $(4,-5)$ satisfies both equations.

$-3x-y = 4 \longrightarrow -3(4)-(-5) = 4 \quad$ No

Therefore, $(4,-5)$ is not a solution of the system.

For Problems 11-30, you need to graph each line, read the coordinates of the point of intersection, and check these coordinates in both equations.

Problem Set 8.6

1. $x+y = 14$
$\underline{x-y = -2}$
$2x \quad\;\; = 12 \quad$ Add the two equations.
$x = 6$

Substitute 6 for x in $x+y = 14$.

$6+y = 14$
$y = 8$

The solution set is $\{(6,8)\}$.

5. $y = 6-x \qquad$ Add x to both sides. $\qquad x+y = \;\; 6$
$x-y = -18 \quad$ Leave alone. $\qquad\qquad \underline{x-y = -18}$
$\qquad\qquad\qquad\qquad\qquad\qquad\qquad\qquad 2x \quad\;\; = -12$
$\qquad\qquad\qquad\qquad\qquad\qquad\qquad\qquad\quad x = -6$

Substitute -6 for x in $y = 6-x$.

$y = 6-(-6) = 12$

The solution set is $\{(-6,12)\}$.

66

9.
$$x + 2y = 5$$
$$\underline{3x - 2y = 6}$$
$$4x = 11 \quad \text{Add the two equations.}$$
$$x = \frac{11}{4}$$

$$x + 2y = 5 \quad \underline{\text{Multiply by } -3.} \quad \rightarrow \quad -3x - 6y = -15$$
$$3x - 2y = 6 \quad \underline{\text{Leave alone.}} \quad \rightarrow \quad \underline{3x - 2y = 6}$$
$$-8y = -9$$
$$y = \frac{9}{8}$$

The solution set is $\{(\frac{11}{4}, \frac{9}{8})\}$.

13.
$$4x + 5y = 9 \quad \underline{\text{Multiply by 6.}} \quad \rightarrow \quad 24x + 30y = 54$$
$$5x - 6y = -50 \quad \underline{\text{Multiply by 5.}} \quad \rightarrow \quad \underline{25x - 30y = -250}$$
$$49x = -196$$
$$x = -4$$

Substitute -4 for x in $4x + 5y = 9$.
$$4(-4) + 5y = 9$$
$$-16 + 5y = 9$$
$$5y = 25$$
$$y = 5$$

The solution set is $\{(-4, 5)\}$.

17.
$$6x + 5y = -6 \quad \underline{\text{Multiply by 3.}} \quad \rightarrow \quad 18x + 15y = -18$$
$$8x - 3y = 21 \quad \underline{\text{Multiply by 5.}} \quad \rightarrow \quad \underline{40x - 15y = 105}$$
$$58x = 87$$
$$x = \frac{87}{58} = \frac{3}{2}$$

Substitute $\frac{3}{2}$ for x in $6x + 5y = -6$.
$$6(\frac{3}{2}) + 5y = -6$$
$$9 + 5y = -6$$
$$5y = -15$$
$$y = -3$$

The solution set is $\{(\frac{3}{2}, -3)\}$.

21.
$$x + y = 750 \quad \underline{\text{Multiply by } -7.} \quad \rightarrow \quad -7x - 7y = -5250$$
$$.07x + .08y = 57.5 \quad \underline{\text{Multiply by 100.}} \quad \rightarrow \quad \underline{7x + 8y = 5750}$$
$$y = 500$$

Substitute 500 for y in $x + y = 750$.
$$x + 500 = 750$$
$$x = 250$$

The solution set is $\{(250, 500)\}$.

25. Let x and y represent the two numbers.

$$\begin{array}{rl}
x+y = 30 & \text{Their sum is 30.} \\
\underline{x-y = 12} & \text{Their difference is 12.} \\
2x \quad = 42 & \\
x = 21 &
\end{array}$$

Substitute 21 for x in x+y = 30.

$$21+y = 30$$
$$y = 9$$

The numbers are 9 and 21.

29. Let x represent the smaller number and y the larger number.

$$\begin{array}{rl}
y = 2x & \text{One number is twice the other.} \\
3x+5y = 78 & \text{The sum of three times the smaller and five times the} \\
& \text{larger is 78.}
\end{array}$$

$$\begin{array}{ll}
2x - y = 0 & \underline{\text{Multiply by 5.}} \\
3x+5y = 78 & \underline{\text{Leave alone.}}
\end{array} \longrightarrow
\begin{array}{rl}
10x-5y = & 0 \\
\underline{3x+5y =} & \underline{78} \\
13x \quad = & 78 \\
x = & 6
\end{array}$$

Substitute 6 for x in y = 2x.

$$y = 2(6) = 12.$$

The numbers are 6 and 12.

33. Let d represent the number of dimes and q the number of quarters.

$$\begin{array}{rl}
d + q = 10 & \text{The total number of coins is 10.} \\
10d+25q = 145 & \text{The amount in cents is 145.}
\end{array}$$

$$\begin{array}{ll}
d + q = 10 & \underline{\text{Multiply by -10.}} \\
10d+25q = 145 & \underline{\text{Leave alone.}}
\end{array} \longrightarrow
\begin{array}{rl}
-10d-10q = & -100 \\
\underline{10d+25q =} & \underline{145} \\
15q = & 45 \\
q = & 3
\end{array}$$

Substitute 3 for q in d+q = 10.

$$d+3 = 10$$
$$d = 7$$

He has 7 dimes and 3 quarters.

37. Let x represent the number of gallons of the 10% solution and y the number of gallons of the 15% solution.

$$\begin{array}{rl}
x + y = 10 & \text{We want a total of 10 gallons.} \\
.10x + .15y = .13(10) & \text{The amount of salt in the 10\% solution plus the amount} \\
& \text{of salt in the 15\% solution equals the total amount of} \\
& \text{salt in the 13\% solution.}
\end{array}$$

$$\begin{array}{ll}
x + y = 10 & \underline{\text{Multiply by -10.}} \\
.10x + .15y = 1.3 & \underline{\text{Multiply by 100.}}
\end{array} \longrightarrow
\begin{array}{rl}
-10x-10y = & -100 \\
\underline{10x+15y =} & \underline{130} \\
5y = & 30 \\
y = & 6
\end{array}$$

Substitute 6 for y in x+y = 10.

$$x+6 = 10$$
$$x = 4$$

We need to mix 4 gallons of the 10% solution with 6 gallons of the 15% solution.

Problem Set 8.7

1. From the first equation, we can substitute 2x-1 for y in the second equation.

$$x+(2x-1) = 14$$
$$3x-1 = 14$$
$$3x = 15$$
$$x = 5$$

Substitute 5 for x in y = 2x-1.

$$y = 2(5)-1 = 9$$

The solution set is $\{(5,9)\}$.

5. From the second equation, we can substitute -2x+7 for y in the first equation.

$$4x-3(-2x+7) = -6$$
$$4x+6x-21 = -6$$
$$10x = 15$$
$$x = \frac{15}{10} = \frac{3}{2}$$

Substitute $\frac{3}{2}$ for x in y = -2x+7.

$$y = -2(\tfrac{3}{2})+7 = 4$$

The solution set is $\{(\frac{3}{2},4)\}$.

9. From the second equation, we can substitute $\frac{3}{4}y$ for x in the first equation.

$$2(\tfrac{3}{4}y) - y = 12$$
$$\tfrac{3}{2}y - y = 12$$
$$\tfrac{1}{2}y = 12$$
$$y = 24$$

Substitute 24 for y in $x = \frac{3}{4}y$.

$$x = \tfrac{3}{4}(24) = 18$$

The solution set is $\{(18,24)\}$.

13. We can equate the values of x from the two equations.

$$4y-1 = 3y+10$$
$$y = 11$$

Substitute 11 for y in x = 3y+10.

$$x = 3(11)+10 = 43$$

The solution set is $\{(43,11)\}$.

17. Solving the second equation for x yields

$$x+5y = -71$$
$$x = -5y-71.$$

Now we can substitute $-5y-71$ for x in the first equation.

$$8(-5y-71)-3y = -9$$
$$-40y-568-3y = -9$$
$$-43y = 559$$
$$y = -13$$

Substitute -13 for y in $x = -5y-71$.

$$x = -5(-13)-71 = -6$$

The solution set is $\{(-6,-13)\}$.

21. Solving the second equation for x yields

$$3x-2y = 0$$
$$3x = 2y$$
$$x = \frac{2}{3}y.$$

Now we can substitute $\frac{2}{3}y$ for x in the first equation.

$$5(\frac{2}{3}y)+7y = 3$$
$$\frac{10}{3}y+7y = 3$$
$$10y+21y = 9$$
$$31y = 9$$
$$y = \frac{9}{31}$$

Substitute $\frac{9}{31}$ for y in $x = \frac{2}{3}y$.

$$x = \frac{2}{3}(\frac{9}{31}) = \frac{6}{31}$$

The solution set is $\{(\frac{6}{31},\frac{9}{31})\}$.

25. Solving the first equation for y produces $y = 13-x$. Now we can substitute $13-x$ for y in the second equation.

$$.05x+.1(13-x) = 1.15$$
$$5x+10(13-x) = 115$$
$$5x+130-10x = 115$$
$$-5x = -15$$
$$x = 3$$

Now we can substitute 3 for x in $x+y = 13$.

$$3+y = 13$$
$$y = 10$$

The solution set is $\{(3,10)\}$.

29. From the second equation, we can substitute $-x$ for y in the first equation.

$$2x+9(-x) = 6$$
$$-7x = 6$$
$$x = -\frac{6}{7}$$

Now we can substitute $-\frac{6}{7}$ for x in $y = -x$.

$$y = -(-\frac{6}{7}) = \frac{6}{7}$$

The solution set is $\{(-\frac{6}{7}, \frac{6}{7})\}$.

33.
| $4x - y = 0$ | Multiply by 2. | $8x-2y = 0$ |
| $7x+2y = 9$ | Leave alone. | $7x+2y = 9$ |

$$15x = 9$$
$$x = \frac{9}{15} = \frac{3}{5}$$

Substitute $\frac{3}{5}$ for x in $4x-y = 0$.

$$4(\frac{3}{5})-y = 0$$
$$\frac{12}{5} = y$$

The solution set is $\{(\frac{3}{5}, \frac{12}{5})\}$.

37.
| $6x - y = -1$ | Multiply by 2. | $12x-2y = -2$ |
| $10x+2y = 13$ | Leave alone. | $10x+2y = 13$ |

$$22x = 11$$
$$x = \frac{11}{22} = \frac{1}{2}$$

Substitute $\frac{1}{2}$ for x in $6x-y = -1$.

$$6(\frac{1}{2})-y = -1$$
$$3-y = -1$$
$$-y = -4$$
$$y = 4$$

The solution set is $\{(\frac{1}{2}, 4)\}$.

41. From the second equation, we can substitute $2y$ for x in the first equation.

$$3(2y)-8y = -5$$
$$6y-8y = -5$$
$$-2y = -5$$
$$y = \frac{5}{2}$$

Substitute $\frac{5}{2}$ for y in $x = 2y$.

$$x = 2(\frac{5}{2}) = 5$$

The solution set is $\{(5, \frac{5}{2})\}$.

45. From the first equation, we can substitute $-y-1$ for x in the second equation.

$$6(-y-1)-5y = 4$$
$$-6y-6-5y = 4$$
$$-11y = 10$$
$$y = -\frac{10}{11}$$

Substitute $-\frac{10}{11}$ for y in $x = -y-1$.

$$x = -(-\frac{10}{11})-1 = \frac{10}{11}-1 = -\frac{1}{11}$$

The solution set is $\{(-\frac{1}{11}, -\frac{10}{11})\}$.

49. Let x and y represent the number of single and double rooms, respectively.

$$x + y = 50 \quad \text{A total of 50 rooms}$$
$$19x+28y = 1265 \quad \text{Total income}$$

$$
\begin{array}{ll}
x + y = 50 & \underline{\text{Multiply by } -19.} \\
19x+28y = 1265 & \underline{\text{Leave alone.}}
\end{array}
\longrightarrow
\begin{array}{l}
-19x-19y = -950 \\
\underline{19x+28y = 1265} \\
9y = 315 \\
y = 35
\end{array}
$$

Substitute 35 for y in $x+y = 50$.

$$x+35 = 50$$
$$x = 15$$

There were 15 single rooms rented and 35 double rooms.

53. Let d and q represent the number of dimes and quarters, respectively.

$$10d+25q = 1205 \quad \text{Total value in cents}$$
$$q = 2d+5 \qquad \text{The number of quarters is 5 more than twice the number of dimes.}$$

From the second equation, we can substitute $2d+5$ for q in the first equation.

$$10d+25(2d+5) = 1205$$
$$10d+50d+125 = 1205$$
$$60d = 1080$$
$$d = 18$$

Substitute 18 for d in $q = 2d+5$.

$$q = 2(18)+5 = 41$$

Larry has 18 dimes and 41 quarters.

57. Let x and y represent the amounts invested at 8% and 9%, respectively.

      $y = x+250$ Invested $250 more at 9% than at 8%
      $.08x + .09y = 48$ Total interest

From the first equation, we can substitute x+250 for y in the second equation.

$$.08x + .09(x+250) = 48$$
$$8x+9(x+250) = 4800$$
$$8x+9x+2250 = 4800$$
$$17x = 2550$$
$$x = 150$$

Substitute 150 for x in $y = x+250$.

   $y = 150+250 = 400$

She invested $150 at 8% and $400 at 9%.

Chapter 9

Problem Set 9.1

1. $\sqrt{49} = 7$ because $7^2 = 49$.   5. $\sqrt{121} = 11$ because $11^2 = 121$.

9. $-\sqrt{1600} = -40$ because $40^2 = 1600$.

13. $\sqrt{19} \approx 4.359$ 17. $\sqrt{75} \approx 8.660$

Locate 19 in   Read this value   Locate 75 in   Read this value
the column   for $\sqrt{19}$ in the   the column   for $\sqrt{75}$ in the
labeled n.   column labeled $\sqrt{n}$.   labeled n.   column labeled $\sqrt{n}$.

21. $\sqrt{196} = 14$ 25. $\sqrt{841} = 29$

Locate 196 in   Read this value   Locate 841 in   Read this value
the column   for $\sqrt{196}$ in the   the column   for $\sqrt{841}$ in the
labeled n^2.   column labeled n.   labeled n^2.   column labeled n.

29. $6^2 = 36$
     $7^2 = 49$ Since 40 is closer to 36 than to 49, we obtain $\sqrt{40} \approx 6$.

33. $34^2 = 1156$
     $35^2 = 1225$ Since 1175 is closer to 1156 than to 1225, we obtain
                         $\sqrt{1175} \approx 34$.

37. $81^2 = 6561$
     $82^2 = 6724$ Since 6614 is closer to 6561 than to 6724, we obtain
                         $\sqrt{6614} \approx 81$.

41. $7\sqrt{2} + 14\sqrt{2} = (7+14)\sqrt{2} = 21\sqrt{2}$ 45. $6\sqrt{3} - 15\sqrt{3} = (6-15)\sqrt{3} = -9\sqrt{3}$

49. $8\sqrt{2} - 4\sqrt{3} - 9\sqrt{2} + 6\sqrt{3} = (8-9)\sqrt{2} + (-4+6)\sqrt{3} = -\sqrt{2} + 2\sqrt{3}$

53. $9\sqrt{3} + \sqrt{3} = (9+1)\sqrt{3} = 10\sqrt{3}$
                         $\approx 10(1.732)$
                         ≈ 17.32
                         $= 17.3$ to the nearest tenth

57. $14\sqrt{2} - 15\sqrt{2} = (14-15)\sqrt{2} = -\sqrt{2}$
                         $\approx -(1.414)$
                         $= -1.4$ to the nearest tenth

61. $4\sqrt{3} - 2\sqrt{2} \approx 4(1.732) - 2(1.414)$
                         $= 4.1$

65. $4\sqrt{11} - 5\sqrt{11} - 7\sqrt{11} + 2\sqrt{11} - 3\sqrt{11} = (4-5-7+2-3)\sqrt{11}$
                         $= -9\sqrt{11}$
                         $\approx -9(3.317)$
                         ≈ -29.853
                         $= -29.9$ to the nearest tenth

Problem Set 9.2

1. $\sqrt{24} = \sqrt{4}\sqrt{6} = 2\sqrt{6}$

5. $\sqrt{27} = \sqrt{9}\sqrt{3} = 3\sqrt{3}$

9. $\sqrt{28} = \sqrt{4}\sqrt{7} = 2\sqrt{7}$

13. $\sqrt{117} = \sqrt{9}\sqrt{13} = 3\sqrt{13}$

17. $3\sqrt{75} = 3\sqrt{25}\sqrt{3} = 3(5)\sqrt{3} = 15\sqrt{3}$

21. $-8\sqrt{96} = -8\sqrt{16}\sqrt{6} = -8(4)\sqrt{6} = -32\sqrt{6}$

25. $\frac{3}{4}\sqrt{12} = \frac{3}{4}\sqrt{4}\sqrt{3} = \frac{3}{4}(2)\sqrt{3} = \frac{3}{2}\sqrt{3}$

29. $-\sqrt{150} = -\sqrt{25}\sqrt{6} = -5\sqrt{6}$

33. $\sqrt{2x^2y} = \sqrt{x^2}\sqrt{2y} = x\sqrt{2y}$

37. $\sqrt{27a^3b} = \sqrt{9a^2}\sqrt{3ab} = 3a\sqrt{3ab}$

41. $\sqrt{63x^4y^2} = \sqrt{9x^4y^2}\sqrt{7} = 3x^2y\sqrt{7}$

45. $-6\sqrt{72x^7} = -6\sqrt{36x^6}\sqrt{2x} = -6(6x^3)\sqrt{2x}$
$$= -36x^3\sqrt{2x}$$

49. $\frac{3}{4}\sqrt{52x^5y^6} = \frac{3}{4}\sqrt{4x^4y^6}\sqrt{13x} = \frac{3}{4}(2x^2y^3)\sqrt{13x} = \frac{3}{2}x^2y^3\sqrt{13x}$

53. $7\sqrt{32} + 5\sqrt{2} = 7\sqrt{16}\sqrt{2} + 5\sqrt{2}$
$$= 28\sqrt{2} + 5\sqrt{2}$$
$$= 33\sqrt{2}$$

57. $3\sqrt{50} + 4\sqrt{18} = 3\sqrt{25}\sqrt{2} + 4\sqrt{9}\sqrt{2}$
$$= 15\sqrt{2} + 12\sqrt{2}$$
$$= 27\sqrt{2}$$

61. $5\sqrt{12} + 3\sqrt{27} - 2\sqrt{75} = 5\sqrt{4}\sqrt{3} + 3\sqrt{9}\sqrt{3} - 2\sqrt{25}\sqrt{3}$
$$= 10\sqrt{3} + 9\sqrt{3} - 10\sqrt{3} = 9\sqrt{3}$$

65. $3\sqrt{8} - 5\sqrt{20} - 7\sqrt{18} - 9\sqrt{125} = 3\sqrt{4}\sqrt{2} - 5\sqrt{4}\sqrt{5} - 7\sqrt{9}\sqrt{2} - 9\sqrt{25}\sqrt{5}$
$$= 6\sqrt{2} - 10\sqrt{5} - 21\sqrt{2} - 45\sqrt{5} = -15\sqrt{2} - 55\sqrt{5}$$

Problem Set 9.3

1. $\sqrt{\frac{16}{25}} = \frac{\sqrt{16}}{\sqrt{25}} = \frac{4}{5}$

5. $\sqrt{\frac{1}{64}} = \frac{\sqrt{1}}{\sqrt{64}} = \frac{1}{8}$

9. $-\sqrt{\frac{25}{256}} = -\frac{\sqrt{25}}{\sqrt{256}} = -\frac{5}{16}$

13. $\sqrt{\frac{8}{49}} = \frac{\sqrt{8}}{\sqrt{49}} = \frac{\sqrt{4}\sqrt{2}}{7} = \frac{2\sqrt{2}}{7}$

17. $\frac{\sqrt{12}}{\sqrt{36}} = \frac{\sqrt{4}\sqrt{3}}{6} = \frac{2\sqrt{3}}{6} = \frac{\sqrt{3}}{3}$

21. $\sqrt{\frac{5}{8}} = \frac{\sqrt{5}}{\sqrt{8}} \cdot \frac{\sqrt{2}}{\sqrt{2}} = \frac{\sqrt{10}}{\sqrt{16}} = \frac{\sqrt{10}}{4}$

25. $\frac{\sqrt{63}}{\sqrt{7}} = \sqrt{\frac{63}{7}} = \sqrt{9} = 3$

29. $\frac{\sqrt{4}}{\sqrt{27}} = \frac{2}{3\sqrt{3}} = \frac{2}{3\sqrt{3}} \cdot \frac{\sqrt{3}}{\sqrt{3}} = \frac{2\sqrt{3}}{9}$

33. $\frac{2\sqrt{3}}{\sqrt{5}} \cdot \frac{\sqrt{5}}{\sqrt{5}} = \frac{2\sqrt{15}}{5}$

37. $\frac{3\sqrt{7}}{4\sqrt{12}} \cdot \frac{\sqrt{3}}{\sqrt{3}} = \frac{3\sqrt{21}}{24} = \frac{\sqrt{21}}{8}$

41. $\frac{3}{\sqrt{x}} \cdot \frac{\sqrt{x}}{\sqrt{x}} = \frac{3\sqrt{x}}{x}$

45. $\sqrt{\frac{3}{x}} = \frac{\sqrt{3}}{\sqrt{x}} \cdot \frac{\sqrt{x}}{\sqrt{x}} = \frac{\sqrt{3x}}{x}$

49. $\frac{\sqrt{2x}}{\sqrt{5y}} \cdot \frac{\sqrt{5y}}{\sqrt{5y}} = \frac{\sqrt{10xy}}{5y}$

53. $\frac{\sqrt{2x^3}}{\sqrt{8y}} = \sqrt{\frac{2x^3}{8y}} = \sqrt{\frac{x^3}{4y}} = \frac{\sqrt{x^3}}{\sqrt{4y}} = \frac{x\sqrt{x}}{2\sqrt{y}}$
$$= \frac{x\sqrt{x}}{2\sqrt{y}} \cdot \frac{\sqrt{y}}{\sqrt{y}} = \frac{x\sqrt{xy}}{2y}$$

75

57. $\dfrac{4}{\sqrt{x^7}} \cdot \dfrac{\sqrt{x}}{\sqrt{x}} = \dfrac{4\sqrt{x}}{\sqrt{x^8}} = \dfrac{4\sqrt{x}}{x^4}$

61. $7\sqrt{3} + \sqrt{\dfrac{1}{3}} = 7\sqrt{3} + \dfrac{1}{\sqrt{3}} \cdot \dfrac{\sqrt{3}}{\sqrt{3}}$

$= 7\sqrt{3} + \dfrac{\sqrt{3}}{3} = \left(7 + \dfrac{1}{3}\right)\sqrt{3}$

$= \dfrac{22}{3}\sqrt{3}$

65. $-2\sqrt{5} - 5\sqrt{\dfrac{1}{5}} = -2\sqrt{5} - \dfrac{5}{\sqrt{5}} = -2\sqrt{5} - \dfrac{5}{\sqrt{5}} \cdot \dfrac{\sqrt{5}}{\sqrt{5}}$

$= -2\sqrt{5} - \sqrt{5} = -3\sqrt{5}$

69. $4\sqrt{12} + \dfrac{3}{\sqrt{3}} - 5\sqrt{27} = 4\sqrt{4}\sqrt{3} + \dfrac{3}{\sqrt{3}} \cdot \dfrac{\sqrt{3}}{\sqrt{3}} - 5\sqrt{9}\sqrt{3}$

$= 8\sqrt{3} + \sqrt{3} - 15\sqrt{3} = -6\sqrt{3}$

Problem Set 9.4

1. $\sqrt{7}\sqrt{5} = \sqrt{35}$

5. $\sqrt{5}\sqrt{10} = \sqrt{50} = \sqrt{25}\sqrt{2} = 5\sqrt{2}$

9. $\sqrt{8}\sqrt{12} = \sqrt{96} = \sqrt{16}\sqrt{6} = 4\sqrt{6}$

13. $(-2\sqrt{2})(3\sqrt{7}) = -6\sqrt{14}$

17. $(5\sqrt{2})(4\sqrt{12}) = 20\sqrt{24} = 20\sqrt{4}\sqrt{6} = 40\sqrt{6}$

21. $\sqrt{2}(\sqrt{3} + \sqrt{5}) = \sqrt{2}\sqrt{3} + \sqrt{2}\sqrt{5} = \sqrt{6} + \sqrt{10}$

25. $\sqrt{3}(\sqrt{6} + \sqrt{7}) = \sqrt{3}\sqrt{6} + \sqrt{3}\sqrt{7} = \sqrt{18} + \sqrt{21} = \sqrt{9}\sqrt{2} + \sqrt{21} = 3\sqrt{2} + \sqrt{21}$

29. $4\sqrt{3}(\sqrt{2} - 2\sqrt{5}) = 4\sqrt{3}(\sqrt{2}) - 4\sqrt{3}(2\sqrt{5}) = 4\sqrt{6} - 8\sqrt{15}$

33. $(\sqrt{6} - 5)(\sqrt{6} + 3) = \sqrt{6}(\sqrt{6} + 3) - 5(\sqrt{6} + 3) = 6 + 3\sqrt{6} - 5\sqrt{6} - 15 = -9 - 2\sqrt{6}$

37. $(5 + \sqrt{10})(5 - \sqrt{10}) = 5^2 - (\sqrt{10})^2 = 25 - 10 = 15$

41. $(5\sqrt{3} + 2\sqrt{6})(5\sqrt{3} - 2\sqrt{6}) = (5\sqrt{3})^2 - (2\sqrt{6})^2 = 75 - 24 = 51$

45. $\sqrt{3x}\sqrt{6y} = \sqrt{18xy} = \sqrt{9}\sqrt{2xy} = 3\sqrt{2xy}$

49. $\sqrt{2x}(\sqrt{3x} - \sqrt{6y}) = \sqrt{2x}(\sqrt{3x}) - \sqrt{2x}(\sqrt{6y}) = \sqrt{6x^2} - \sqrt{12xy} = x\sqrt{6} - 2\sqrt{3xy}$

53. $(\sqrt{x} + 7)(\sqrt{x} - 7) = (\sqrt{x})^2 - 7^2 = x - 49$

57. $\dfrac{8}{\sqrt{6} - 2} \cdot \dfrac{\sqrt{6} + 2}{\sqrt{6} + 2} = \dfrac{8(\sqrt{6} + 2)}{6 - 4} = \dfrac{8(\sqrt{6} + 2)}{2} = 4(\sqrt{6} + 2) = 4\sqrt{6} + 8$

61. $\dfrac{10}{2 - 3\sqrt{3}} \cdot \dfrac{2 + 3\sqrt{3}}{2 + 3\sqrt{3}} = \dfrac{10(2 + 3\sqrt{3})}{4 - 27} = \dfrac{10(2 + 3\sqrt{3})}{-23} = \dfrac{-20 - 30\sqrt{3}}{23}$

65. $\dfrac{\sqrt{x}}{\sqrt{x} + 3} \cdot \dfrac{\sqrt{x} - 3}{\sqrt{x} - 3} = \dfrac{\sqrt{x}(\sqrt{x} - 3)}{x - 9} = \dfrac{x - 3\sqrt{x}}{x - 9}$

69. $\dfrac{2 + \sqrt{3}}{3 - \sqrt{2}} \cdot \dfrac{3 + \sqrt{2}}{3 + \sqrt{2}} = \dfrac{(2 + \sqrt{3})(3 + \sqrt{2})}{9 - 2} = \dfrac{6 + 2\sqrt{2} + 3\sqrt{3} + \sqrt{6}}{7}$

Problem Set 9.5

1. $\sqrt{x} = 7$ Check

      $(\sqrt{x})^2 = 7^2$ $\sqrt{49} \overset{?}{=} 7$

         $x = 49$ $7 = 7$

The solution set is $\{49\}$.

5. $\sqrt{3x} = -6$ The solution set is $\emptyset$ because $\sqrt{3x}$ will always be non-negative.

9. $3\sqrt{x} = 2$ Check

    $(3\sqrt{x})^2 = 2^2$ $3\sqrt{\dfrac{4}{9}} \overset{?}{=} 2$

       $9x = 4$ $3\left(\dfrac{2}{3}\right) \overset{?}{=} 2$

          $x = \dfrac{4}{9}$ $2 = 2$

The solution set is $\{\dfrac{4}{9}\}$.

13. $\sqrt{5y+2} = -1$ The solution set is $\emptyset$ because $\sqrt{5y+2}$ will always be non-negative.

17. $5\sqrt{x} = 30$ Check

        $\sqrt{x} = 6$ $5\sqrt{36} \overset{?}{=} 30$

    $(\sqrt{x})^2 = 6^2$ $5(6) = 30$

        $x = 36$

The solution set is $\{36\}$.

21. $\sqrt{7x-3} = \sqrt{4x-3}$ Check

    $(\sqrt{7x-3})^2 = (\sqrt{4x+3})^2$ $\sqrt{7(2)-3} \overset{?}{=} \sqrt{4(2)+3}$

        $7x-3 = 4x+3$ $\sqrt{11} = \sqrt{11}$

          $3x = 6$

            $x = 2$

        The solution set is $\{2\}$.

25. $\sqrt{x+3} = x+3$

    $(\sqrt{x+3})^2 = (x+3)^2$

        $x+3 = x^2+6x+9$

          $0 = x^2+5x+6$

          $0 = (x+3)(x+2)$

    $x+3 = 0$ or $x+2 = 0$

      $x = -3$ or $x = -2$

                 Check

    $\sqrt{-3+3} \overset{?}{=} -3+3$ $\sqrt{-2+3} \overset{?}{=} -2+3$

        $0 = 0$ $\sqrt{1} \overset{?}{=} 1$

                            $1 = 1$

The solution set is $\{-3,-2\}$.

29. $\sqrt{3n-4} = \sqrt{n}$

       $(\sqrt{3n-4})^2 = (\sqrt{n})^2$

          $3n-4 = n$

            $2n = 4$

             $n = 2$

Check

$\sqrt{3(2)-4} \stackrel{?}{=} \sqrt{2}$

$\sqrt{2} = \sqrt{2}$

The solution set is $\{2\}$.

33. $4\sqrt{x} + 5 = x$

        $4\sqrt{x} = x-5$

      $(4\sqrt{x})^2 = (x-5)^2$

          $16x = x^2-10x+25$

           $0 = x^2-26x+25$

           $0 = (x-25)(x-1)$

   $x-25 = 0$ or $x-1 = 0$

      $x = 25$ or $x = 1$

Check

$4\sqrt{25} + 5 \stackrel{?}{=} 25$ $4\sqrt{1} + 5 \stackrel{?}{=} 1$

$20+5 = 25$ $9 \neq 1$

The solution set is $\{25\}$.

37. $\sqrt{x^2+2x+3} = x+2$

    $\left(\sqrt{x^2+2x+3}\right)^2 = (x+2)^2$

       $x^2+2x+3 = x^2+4x+4$

            $-1 = 2x$

          $-\dfrac{1}{2} = x$

Check

$\sqrt{\left(-\dfrac{1}{2}\right)^2 + 2\left(-\dfrac{1}{2}\right) + 3} \stackrel{?}{=} -\dfrac{1}{2} + 2$

$\sqrt{\dfrac{1}{4} - 1 + 3} \stackrel{?}{=} \dfrac{3}{2}$

$\sqrt{\dfrac{9}{4}} \stackrel{?}{=} \dfrac{3}{2}$

$\dfrac{3}{2} = \dfrac{3}{2}$

The solution set is $\{-\dfrac{1}{2}\}$.

Chapter 10

Problem Set 10.1

1. $x^2+15x = 0$
   $x(x+15) = 0$
   $x = 0$ or $x+15 = 0$
   $x = 0$ or $x = -15$

   The solution set is $\{-15,0\}$.

5. $3y^2 = 15y$
   $y^2 = 5y$
   $y^2-5y = 0$
   $y(y-5) = 0$
   $y = 0$ or $y-5 = 0$
   $y = 0$ or $y = 5$

   The solution set is $\{0,5\}$.

9. $x^2-5x-14 = 0$
   $(x-7)(x+2) = 0$
   $x-7 = 0$ or $x+2 = 0$
   $x = 7$ or $x = -2$

   The solution set is $\{-2,7\}$.

13. $6y^2+7y-5 = 0$
    $(3y+5)(2y-1) = 0$
    $3y+5 = 0$ or $2y-1 = 0$
    $3y = -5$ or $2y = 1$
    $y = -\frac{5}{3}$ or $y = \frac{1}{2}$

    The solution set is $\{-\frac{5}{3},\frac{1}{2}\}$.

17. $4x^2-4x+1 = 0$
    $(2x-1)(2x-1) = 0$
    $2x-1 = 0$
    $2x = 1$
    $x = \frac{1}{2}$

    The solution set is $\{\frac{1}{2}\}$.

21. $x^2 = \frac{25}{9}$
    $x = \pm\frac{5}{3}$

    The solution set is $\{-\frac{5}{3},\frac{5}{3}\}$.

25. $n^2 = 14$
    $n = \pm\sqrt{14}$
    The solution set is $\{-\sqrt{14}, \sqrt{14}\}$.

29. $y^2 = 32$
    $y = \pm\sqrt{32} = \pm4\sqrt{2}$
    The solution set is $\{-4\sqrt{2}, 4\sqrt{2}\}$.

33. $2x^2 = 9$
    $x^2 = \frac{9}{2}$
    $x = \pm\sqrt{\frac{9}{2}} = \pm\frac{\sqrt{9}}{\sqrt{2}} = \pm\frac{3}{\sqrt{2}}\cdot\frac{\sqrt{2}}{\sqrt{2}} = \pm\frac{3\sqrt{2}}{2}$

    The solution set is $\{-\frac{3\sqrt{2}}{2}, \frac{3\sqrt{2}}{2}\}$.

37. $(x-1)^2 = 4$
    $x-1 = \pm 2$
    $x-1 = -2$ or $x-1 = 2$
    $x = -1$ or $x = 3$

    The solution set is $\{-1,3\}$.

41. $(3x-2)^2 = 49$
    $3x-2 = \pm 7$
    $3x-2 = -7$ or $3x-2 = 7$
    $3x = -5$ or $3x = 9$
    $x = -\frac{5}{3}$ or $x = 3$

    The solution set is $\{-\frac{5}{3},3\}$.

45. $(n-1)^2 = 8$

$n-1 = \pm\sqrt{8} = \pm 2\sqrt{2}$

$n-1 = -2\sqrt{2}$ or $n-1 = 2\sqrt{2}$

$n = 1-2\sqrt{2}$ or $n = 1+2\sqrt{2}$

The solution set is $\{1-2\sqrt{2},\ 1+2\sqrt{2}\}$.

49. $(4x-1)^2 = -2$ The solution set is $\emptyset$ because $(4x-1)^2$ will always be nonnegative.

53. $2(7x-1)^2+5 = 37$

$2(7x-1)^2 = 32$

$(7x-1)^2 = 16$

$7x-1 = \pm 4$

$7x-1 = -4$ or $7x-1 = 4$

$7x = -3$ or $7x = 5$

$x = -\dfrac{3}{7}$ or $x = \dfrac{5}{7}$

The solution set is $\{-\dfrac{3}{7}, \dfrac{5}{7}\}$.

57. $c^2 = a^2+b^2$

$c^2 = 1^2+7^2 = 1+49 = 50$

$c = \sqrt{50} = 5\sqrt{2}$ inches

61. $c^2 = a^2+b^2$

$12^2 = 10^2+b^2$

$144 = 100+b^2$

$44 = b^2$

$\sqrt{44} = b$

$b = 2\sqrt{11}$ feet

65.

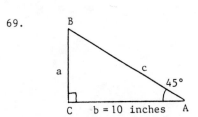

If $a = 6$ feet, then $c = 2(6) = 12$ feet.

$a^2+b^2 = c^2$

$6^2+b^2 = 12^2$

$36+b^2 = 144$

$b^2 = 108$

$b = \sqrt{108} = \sqrt{36}\sqrt{3}$

$= 6\sqrt{3}$ feet

69.

If $b = 10$ inches, then $a = 10$ inches.

$c^2 = a^2+b^2$

$c^2 = 10^2+10^2$

$c^2 = 100+100$

$c^2 = 200$

$c = \sqrt{200} = \sqrt{100}\sqrt{2} = 10\sqrt{2}$ inches

Problem Set 10.2

1. $x^2+8x-1 = 0$

   $x^2+8x = 1$

   $x^2+8x+16 = 1+16$

   $(x+4)^2 = 17$

   $x+4 = \pm\sqrt{17}$

   $x+4 = -\sqrt{17}$ or $x+4 = \sqrt{17}$

   $x = -4-\sqrt{17}$ or $x = -4+\sqrt{17}$

The solution set is $\{-4-\sqrt{17}, -4+\sqrt{17}\}$.

5. $x^2-4x-4 = 0$

   $x^2-4x = 4$

   $x^2-4x+4 = 4+4$

   $(x-2)^2 = 8$

   $x-2 = \pm\sqrt{8} = \pm2\sqrt{2}$

   $x-2 = -2\sqrt{2}$ or $x-2 = 2\sqrt{2}$

   $x = 2-2\sqrt{2}$ or $x = 2+2\sqrt{2}$

The solution set is $\{2-2\sqrt{2}, 2+2\sqrt{2}\}$.

9. $n^2+2n = 17$

   $n^2+2n+1 = 17+1$

   $(n+1)^2 = 18$

   $n+1 = \pm\sqrt{18} = \pm3\sqrt{2}$

   $n+1 = -3\sqrt{2}$ or $n+1 = 3\sqrt{2}$

   $n = -1-3\sqrt{2}$ or $n = -1+3\sqrt{2}$

The solution set is $\{-1-3\sqrt{2}, -1+3\sqrt{2}\}$.

13. $a^2-5a = 2$

   $a^2-5a+\frac{25}{4} = 2+\frac{25}{4}$

   $\left(a-\frac{5}{2}\right)^2 = \frac{33}{4}$

   $a-\frac{5}{2} = \pm\frac{\sqrt{33}}{2}$

   $a-\frac{5}{2} = -\frac{\sqrt{33}}{2}$ or $a-\frac{5}{2} = \frac{\sqrt{33}}{2}$

   $a = \frac{5}{2}-\frac{\sqrt{33}}{2}$ or $a = \frac{5}{2}+\frac{\sqrt{33}}{2}$

   $a = \frac{5-\sqrt{33}}{2}$ or $a = \frac{5+\sqrt{33}}{2}$

The solution set is $\{\frac{5-\sqrt{33}}{2}, \frac{5+\sqrt{33}}{2}\}$.

17. $3x^2+12x-2 = 0$

   $3x^2+12x = 2$

   $x^2+4x = \frac{2}{3}$

   $x^2+4x+4 = \frac{2}{3}+4$

   $(x+2)^2 = \frac{14}{3}$

   $x+2 = \pm\frac{\sqrt{14}}{3} = \pm\frac{\sqrt{42}}{3}$

   $x+2 = -\frac{\sqrt{42}}{3}$ or $x+2 = \frac{\sqrt{42}}{3}$

   $x = -2-\frac{\sqrt{42}}{3}$ or $x = -2+\frac{\sqrt{42}}{3}$

   $x = \frac{-6-\sqrt{42}}{3}$ or $x = \frac{-6+\sqrt{42}}{3}$

The solution set is $\{\frac{-6-\sqrt{42}}{3}, \frac{-6+\sqrt{42}}{3}\}$.

21. $5n^2+10n+6 = 0$

   $5n^2+10n = -6$

   $n^2+2n = -\frac{6}{5}$

   $n^2+2n+1 = -\frac{6}{5}+1$

   $(n+1)^2 = -\frac{1}{5}$

At this stage we can see that the solution set is $\emptyset$ because $(n+1)^2$ will always be nonnegative.

25. $2x^2+3x-1 = 0$

$\qquad 2x^2+3x = 1$

$\qquad x^2 + \frac{3}{2}x = \frac{1}{2}$

$\qquad x^2 + \frac{3}{2}x + \frac{9}{16} = \frac{1}{2} + \frac{9}{16}$

$\qquad \left(x + \frac{3}{4}\right)^2 = \frac{17}{16}$

$\qquad x + \frac{3}{4} = \pm \frac{\sqrt{17}}{4}$

$x + \frac{3}{4} = -\frac{\sqrt{17}}{4}$ or $x + \frac{3}{4} = \frac{\sqrt{17}}{4}$

$x = -\frac{3}{4} - \frac{\sqrt{17}}{4}$ or $x = -\frac{3}{4} + \frac{\sqrt{17}}{4}$

$x = \frac{-3 - \sqrt{17}}{4}$ or $x = \frac{-3 + \sqrt{17}}{4}$

The solution set is $\left\{ \frac{-3-\sqrt{17}}{4}, \frac{-3+\sqrt{17}}{4} \right\}$.

29. $n(n+2) = 168$

$\qquad n^2+2n = 168$

$\qquad n^2+2n+1 = 168+1$

$\qquad (n+1)^2 = 169$

$\qquad n+1 = \pm 13$

$n+1 = -13$ or $n+1 = 13$

$n = -14$ or $n = 12$

The solution set is $\{-14, 12\}$.

33.

Factoring	Completing the Square
$x^2+4x-12 = 0$	$x^2+4x+4 = 12+4$
$(x+6)(x-2) = 0$	$(x+2)^2 = 16$
$x+6 = 0$ or $x-2 = 0$	$x+2 = \pm 4$
$x = -6$ or $x = 2$	$x+2 = -4$ or $x+2 = 4$
	$x = -6$ or $x = 2$

The solution set is $\{-6,2\}$.

37.

Factoring	Completing the Square
$n^2-3n-40 = 0$	$n^2-3n + \frac{9}{4} = 40 + \frac{9}{4}$
$(n-8)(n+5) = 0$	$\left(n - \frac{3}{2}\right)^2 = \frac{169}{4}$
$n-8 = 0$ or $n+5 = 0$	$n - \frac{3}{2} = \pm \frac{13}{2}$
$n = 8$ or $n = -5$	
	$n - \frac{3}{2} = -\frac{13}{2}$ or $n - \frac{3}{2} = \frac{13}{2}$
	$n = -5$ or $n = 8$

The solution set is $\{-5,8\}$.

41.

Factoring	Completing the Square

Factoring

$$4n^2+4n-15 = 0$$
$$(2n-3)(2n+5) = 0$$
$$2n-3 = 0 \text{ or } 2n+5 = 0$$
$$2n = 3 \text{ or } \quad 2n = -5$$
$$n = \frac{3}{2} \text{ or } \quad n = -\frac{5}{2}$$

Completing the Square

$$4n^2+4n = 15$$
$$n^2+n = \frac{15}{4}$$
$$n^2+n+\frac{1}{4} = \frac{15}{4}+\frac{1}{4}$$
$$\left(n+\frac{1}{2}\right)^2 = 4$$
$$n+\frac{1}{2} = \pm 2$$
$$n+\frac{1}{2} = -2 \text{ or } n+\frac{1}{2} = 2$$
$$n = -\frac{5}{2} \text{ or } \quad n = \frac{3}{2}$$

The solution set is $\{-\frac{5}{2},\frac{3}{2}\}$.

Problem Set 10.3

1. $x^2-5x-6 = 0$

$$x = \frac{-(-5) \pm \sqrt{(-5)^2-4(1)(-6)}}{2(1)}$$
$$x = \frac{5 \pm \sqrt{49}}{2}$$
$$x = \frac{5 \pm 7}{2}$$
$$x = \frac{5-7}{2} \text{ or } x = \frac{5+7}{2}$$
$$x = -1 \text{ or } x = 6$$

The solution set is $\{-1,6\}$.

9. $x^2-2x+6 = 0$

$$x = \frac{-(-2) \pm \sqrt{(-2)^2-4(1)(6)}}{2(1)}$$
$$x = \frac{2 \pm \sqrt{-20}}{2}$$

Since $\sqrt{-20}$ is not a real number, this equation has no real number solutions. Its solution set is $\emptyset$.

5. $n^2-2n-5 = 0$

$$n = \frac{-(-2) \pm \sqrt{(-2)^2-4(1)(-5)}}{2(1)}$$
$$n = \frac{2 \pm \sqrt{24}}{2} = \frac{2 \pm 2\sqrt{6}}{2} = 1 \pm \sqrt{6}$$

The solution set is $\{1-\sqrt{6},\ 1+\sqrt{6}\}$.

13. $x^2-6x = 0$

$$x = \frac{-(-6) \pm \sqrt{(-6)^2-4(1)(0)}}{2(1)}$$
$$x = \frac{6 \pm \sqrt{36}}{2} = \frac{6 \pm 6}{2}$$
$$x = \frac{6-6}{2} \text{ or } x = \frac{6+6}{2}$$
$$x = 0 \quad \text{ or } x = 6$$

The solution set is $\{0,6\}$.

17. $n^2-34n+288 = 0$

$$n = \frac{-(-34) \pm \sqrt{(-34)^2-4(1)(288)}}{2(1)}$$

$$n = \frac{34 \pm \sqrt{4}}{2} = \frac{34 \pm 2}{2}$$

$$n = \frac{34-2}{2} \text{ or } n = \frac{34+2}{2}$$

$$n = 16 \quad \text{ or } n = 18$$

The solution set is $\{16,18\}$.

21. $t^2+4t+4 = 0$

$$t = \frac{-4 \pm \sqrt{4^2-4(1)(4)}}{2(1)}$$

$$t = \frac{-4 \pm \sqrt{0}}{2} = -2$$

The solution set is $\{-2\}$.

25. $5x^2+3x-2 = 0$

$$x = \frac{-3 \pm \sqrt{3^2-4(5)(-2)}}{2(5)}$$

$$x = \frac{-3 \pm \sqrt{49}}{10} = \frac{-3 \pm 7}{10}$$

$$x = \frac{-3-7}{10} \text{ or } x = \frac{-3+7}{10}$$

$$x = -1 \quad \text{ or } x = \frac{2}{5}$$

The solution set is $\{-1, \frac{2}{5}\}$.

29. $2x^2+5x-6 = 0$

$$x = \frac{-5 \pm \sqrt{5^2-4(2)(-6)}}{2(2)}$$

$$x = \frac{-5 \pm \sqrt{73}}{4}$$

The solution set is $\{\frac{-5 - \sqrt{73}}{4}, \frac{-5 + \sqrt{73}}{4}\}$.

33. $16x^2+24x+9 = 0$

$$x = \frac{-24 \pm \sqrt{24^2-4(16)(9)}}{2(16)}$$

$$x = \frac{-24 \pm \sqrt{0}}{32} = -\frac{3}{4}$$

The solution set is $\{-\frac{3}{4}\}$.

37. $6n^2+9n+1 = 0$

$$n = \frac{-9 \pm \sqrt{9^2-4(6)(1)}}{2(6)}$$

$$n = \frac{-9 \pm \sqrt{57}}{12}$$

The solution set is $\{\frac{-9 - \sqrt{57}}{12}, \frac{-9 + \sqrt{57}}{12}\}$.

41. $4t^2+5t+3 = 0$

$$t = \frac{-5 \pm \sqrt{25-4(4)(3)}}{2(4)}$$

$$t = \frac{-5 \pm \sqrt{-23}}{8}$$

The solution set is $\emptyset$ because $\sqrt{-23}$ is not a real number.

45. $7 = 3x^2-x$

$0 = 3x^2-x-7$

$$x = \frac{-(-1) \pm \sqrt{(-1)^2-4(3)(-7)}}{2(3)}$$

$$x = \frac{1 \pm \sqrt{85}}{6}$$

The solution set is $\{\frac{1 - \sqrt{85}}{6}, \frac{1 + \sqrt{85}}{6}\}$.

Problem Set 10.4

For the problems in this set, I have shown the method that I would use to solve the particular quadratic equation. You may choose a different method, but we should end up with the same solutions.

1. $x^2+4x = 45$
$$x^2+4x-45 = 0$$
$$(x+9)(x-5) = 0$$
$$x+9 = 0 \quad \text{or} \quad x-5 = 0$$
$$x = -9 \text{ or} \quad x = 5$$

    The solution set is $\{-9,5\}$.

5. $t^2-t-2 = 0$
$$(t-2)(t+1) = 0$$
$$t-2 = 0 \text{ or } t+1 = 0$$
$$t = 2 \text{ or} \quad t = -1$$

    The solution set is $\{-1,2\}$.

9. $9x^2-6x+1 = 0$
$$(3x-1)^2 = 0$$
$$3x-1 = 0$$
$$3x = 1$$
$$x = \frac{1}{3}$$

    The solution set is $\{\frac{1}{3}\}$.

13. $n^2-14n = 19$
$$n^2-14n+49 = 19+49$$
$$(n-7)^2 = 68$$
$$n-7 = \pm\sqrt{68} = \pm 2\sqrt{17}$$
$$n-7 = -2\sqrt{17} \quad \text{or } n-7 = 2\sqrt{17}$$
$$n = 7-2\sqrt{17} \text{ or} \quad n = 7+2\sqrt{17}$$

The solution set is $\{7-2\sqrt{17},\ 7+2\sqrt{17}\}$.

17. $15x^2+28x+5 = 0$
$$(5x+1)(3x+5) = 0$$
$$5x+1 = 0 \quad \text{or } 3x+5 = 0$$
$$5x = -1 \text{ or} \quad 3x = -5$$
$$x = -\frac{1}{5} \text{ or} \quad x = -\frac{5}{3}$$

    The solution set is $\{-\frac{5}{3}, -\frac{1}{5}\}$.

21. $y^2+5y = 84$
$$y^2+5y-84 = 0$$
$$(y+12)(y-7) = 0$$
$$y+12 = 0 \quad \text{or } y-7 = 0$$
$$y = -12 \text{ or} \quad y = 7$$

    The solution set is $\{-12,7\}$.

25. $3x^2-9x-12 = 0$
$$x^2-3x-4 = 0$$
$$(x-4)(x+1) = 0$$
$$x-4 = 0 \text{ or } x+1 = 0$$
$$x = 4 \text{ or} \quad x = -1$$

    The solution set is $\{-1,4\}$.

29. $n(n-46) = -480$
$$n^2-46n = -480$$
$$n^2-46n+529 = -480+529$$
$$(n-23)^2 = 49$$
$$n-23 = \pm 7$$
$$n-23 = -7 \text{ or } n-23 = 7$$
$$n = 16 \text{ or} \quad n = 30$$

    The solution set is $\{16,30\}$.

33. $x + \frac{1}{x} = \frac{25}{12}$
$$12x^2+12 = 25x$$
$$12x^2-25x+12 = 0$$
$$(4x-3)(3x-4) = 0$$
$$4x-3 = 0 \text{ or } 3x-4 = 0$$
$$4x = 3 \text{ or} \quad 3x = 4$$
$$x = \frac{3}{4} \text{ or} \quad x = \frac{4}{3}$$

    The solution set is $\{\frac{3}{4}, \frac{4}{3}\}$.

37. $x^2-28x+187 = 0$
$$(x-11)(x-17) = 0$$
$$x-11 = 0 \quad \text{or } x-17 = 0$$
$$x = 11 \text{ or} \quad x = 17$$

    The solution set is $\{11,17\}$.

41.
$$\frac{2}{x+2} - \frac{1}{x} = 3$$

$$x(x+2)\left[\frac{2}{x+2} - \frac{1}{x}\right] = 3x(x+2)$$

$$2x-(x+2) = 3x^2+6x$$

$$2x-x-2 = 3x^2+6x$$

$$0 = 3x^2+5x+2$$

$$0 = (3x+2)(x+1)$$

$$3x+2 = 0 \quad \text{or} \quad x+1 = 0$$

$$3x = -2 \quad \text{or} \quad x = -1$$

$$x = -\frac{2}{3} \quad \text{or} \quad x = -1$$

The solution set is $\{-1, -\frac{2}{3}\}$.

45.
$$(n-2)(n+4) = 7$$

$$n^2+2n-8 = 7$$

$$n^2+2n-15 = 0$$

$$(n+5)(n-3) = 0$$

$$n+5 = 0 \quad \text{or} \quad n-3 = 0$$

$$n = -5 \quad \text{or} \quad n = 3$$

The solution set is $\{-5, 3\}$.

Problem Set 10.5

1. Let n and n+1 represent the consecutive whole numbers.

$$n(n+1) = 306$$

$$n^2+n-306 = 0$$

$$(n+18)(n-17) = 0$$

$$n+18 = 0 \quad \text{or} \quad n-17 = 0$$

$$n = -18 \quad \text{or} \quad n = 17$$

The negative solution must be discarded since we are looking for whole numbers. The numbers are 17 and 17+1 = 18.

5. Let x and y represent the two numbers.

$$x+y = 6 \qquad \text{Their sum is 6.}$$
$$xy = 4 \qquad \text{Their product is 4.}$$

From the first equation, we obtain y = 6-x. Substitute 6-x for y in the second equation.

$$x(6-x) = 4$$

$$6x-x^2 = 4$$

$$0 = x^2-6x+4$$

$$x = \frac{-(-6)+ \sqrt{(-6)^2-4(1)(4)}}{2(1)}$$

$$x = \frac{6 \pm \sqrt{20}}{2} = \frac{6 \pm 2\sqrt{5}}{2} = 3 \pm \sqrt{5}$$

The numbers are $3 - \sqrt{5}$ and $3 + \sqrt{5}$.

9. Let n, n+2, and n+4 represent the three consecutive even whole numbers.

$$n^2+(n+2)^2+(n+4)^2 = 596$$
$$n^2+n^2+4n+4+n^2+8n+16 = 596$$
$$3n^2+12n+20 = 596$$
$$3n^2+12n-576 = 0$$
$$n^2+4n-192 = 0$$
$$(n+16)(n-12) = 0$$
$$n+16 = 0 \quad \text{or} \quad n-12 = 0$$
$$n = -16 \quad \text{or} \quad n = 12$$

The negative solution must be discarded since we are looking for whole numbers. Therefore, the numbers are 12, 12+2 = 14, and 12+4 = 16.

13. Let w represent the width and 2w-4 the length.

$$w(2w-4) = 96$$
$$2w^2-4w = 96$$
$$2w^2-4w-96 = 0$$
$$w^2-2w-48 = 0$$
$$(w-8)(w+6) = 0$$
$$w-8 = 0 \quad \text{or} \quad w+6 = 0$$
$$w = 8 \quad \text{or} \quad w = -6$$

The negative solution must be discarded because we are dealing with lengths of line segments. Therefore, the width is 8 meters and the length is 2(8)-4 = 12 meters.

17. Let w represent the width and $\frac{26}{9}w$ the length.

$$w(\frac{26}{9}w) = 2106$$
$$26w^2 = 9(2106) = 18954$$
$$w^2 = 729$$
$$w = \pm 27$$

The negative solution must be discarded. Therefore, the width is 27 feet and the length is $\frac{26}{9}(27) = 78$ feet.

21. Let ℓ and w represent the length and width, respectively, of the original rectangle.

$$\ell w = 63 \qquad \text{The area is 63 square feet.}$$
$$(\ell+3)(w+3) = 63+57 = 120 \qquad \text{If the length and width are each increased by 3, the area is increased by 57.}$$

From the first equation we obtain $w = \frac{63}{\ell}$. Substitute $\frac{63}{\ell}$ for w in the second equation.

$$(\ell+3)(\frac{63}{\ell} + 3) = 120$$
$$63 + 3\ell + \frac{189}{\ell} + 9 = 120$$
$$63\ell + 3\ell^2 + 189 + 9\ell = 120$$
$$3\ell^2 - 48\ell + 189 = 0$$
$$\ell^2 - 16\ell + 63 = 0$$
$$(\ell-7)(\ell-9) = 0$$

$\ell-7 = 0$ or $\ell-9 = 0$

$\ell = 7$ or $\ell = 9$

The rectangle is 7 feet by 9 feet.

25.

Let x represent the width of the frame.

$(5+2x)(7+2x) = 80$

$35+24x+4x^2 = 80$

$4x^2+24x-45 = 0$

$(2x+15)(2x-3) = 0$

$2x+15 = 0$ or $2x-3 = 0$

$2x = -15$ or $2x = 3$

$x = -\dfrac{15}{2}$ or $x = \dfrac{3}{2}$

The negative solution must be discarded. Therefore, the width of the frame is $1\frac{1}{2}$ inches.

29. Let x and 56-x represent the lengths of the two pieces. If the piece of length x is bent into a square, the length of each side of the square is $\frac{x}{4}$. If the piece of length 56-x is bent into a square, the length of each side is $\frac{56-x}{4}$.

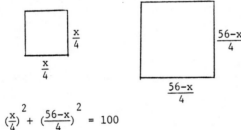

$(\frac{x}{4})^2 + (\frac{56-x}{4})^2 = 100$

$x^2+3136-112x+x^2 = 1600$

$2x^2-112x+1536 = 0$

$x^2-56x+768 = 0$

$(x-24)(x-32) = 0$

$x-24 = 0$ or $x-32 = 0$

$x = 24$ or $x = 32$

The pieces are of length 24 inches and 32 inches.

Chapter 11

Problem Set 11.1

1. $\sqrt{-64} = i\sqrt{64} = 8i$

5. $\sqrt{-11} = i\sqrt{11}$

9. $\sqrt{-48} = i\sqrt{48} = i\sqrt{16}\sqrt{3} = 4i\sqrt{3}$

13. $(3+8i)+(5+9i) = (3+5)+(8+9)i = 8+17i$

17. $(10+4i)-(6+2i) = (10-6)+(4-2)i = 4+2i$

21. $(-2-i)-(3-4i) = (-2-3)+(-1+4)i = -5+3i$

25. $(0-6i)+(-10+2i) = (0+(-10))+(-6+2)i = -10-4i$

29. $(-10-4i)-(10+4i) = (-10-10)+(-4-4)i = -20-8i$

33. $(2i)(6+3i) = 12i+6i^2 = 12i+6(-1) = -6+12i$

37. $(2+3i)(5+4i) = 2(5+4i)+3i(5+4i) = 10+8i+15i+12i^2 = 10+23i+12(-1) = -2+23i$

41. $(-2-3i)(6-3i) = -2(6-3i)-3i(6-3i) = -12+6i-18i+9i^2 = -12-12i+9(-1) = -21-12i$

45. $(4+5i)^2 = 16+40i+25i^2 = 16+40i+25(-1) = -9+40i$

49. $(-2+i)(-2-i) = 4+2i-2i-i^2 = 4-(-1) = 5 = 5+0i$

Problem Set 11.2

1. $x^2 = -64$

$x = \pm\sqrt{-64} = \pm i\sqrt{64} = \pm 8i$

The solution set is $\{-8i, 8i\}$.

5. $(x+5)^2 = -13$

$x+5 = \pm\sqrt{-13} = \pm i\sqrt{13}$

$x+5 = -i\sqrt{13}$ or $x+5 = i\sqrt{13}$

$x = -5 - i\sqrt{13}$ or $x = -5 + i\sqrt{13}$

The solution set is $\{-5 - i\sqrt{13}, -5 + i\sqrt{13}\}$.

9. $(5x-1)^2 = 9$

$5x-1 = \pm\sqrt{9} = \pm 3$

$5x-1 = -3$ or $5x-1 = 3$

$5x = -2$ or $5x = 4$

$x = -\dfrac{2}{5}$ or $x = \dfrac{4}{5}$

The solution set is $\{-\dfrac{2}{5}, \dfrac{4}{5}\}$.

13. $t^2+6t = -12$

$t^2+6t+9 = -12+9$

$(t+3)^2 = -3$

$t+3 = \pm\sqrt{-3} = \pm i\sqrt{3}$

$t+3 = -i\sqrt{3}$ or $t+3 = i\sqrt{3}$

$t = -3 - i\sqrt{3}$ or $t = -3 + i\sqrt{3}$

The solution set is $\{-3 - i\sqrt{3}, -3 + i\sqrt{3}\}$.

17. $x^2-4x+20 = 0$

$$x = \frac{-(-4) \pm \sqrt{(-4)^2-4(1)(20)}}{2(1)}$$

$$x = \frac{4 \pm \sqrt{-64}}{2} = \frac{4 \pm 8i}{2} = 2 \pm 4i$$

The solution set is $\{2-4i, 2+4i\}$.

21. $2x^2-3x-5 = 0$

$(2x-5)(x+1) = 0$

$2x-5 = 0$ or $x+1 = 0$

$2x = 5$ or $x = -1$

$x = \frac{5}{2}$ or $x = -1$

The solution set is $\{-1, \frac{5}{2}\}$.

25. $x^2-4x+7 = 0$

$$x = \frac{4 \pm \sqrt{(-4)^2-4(1)(7)}}{2(1)}$$

$$x = \frac{4 \pm \sqrt{-12}}{2} = \frac{4 \pm 2i\sqrt{3}}{2} = 2 \pm i\sqrt{3}$$

The solution set is $\{2 - i\sqrt{3}, 2 + i\sqrt{3}\}$.

29. $6x^2+2x+1 = 0$

$$x = \frac{-2 \pm \sqrt{2^2-4(6)(1)}}{2(6)}$$

$$x = \frac{-2 \pm \sqrt{-20}}{12} = \frac{-2 \pm 2i\sqrt{5}}{12} = \frac{-1 \pm i\sqrt{5}}{6}$$

The solution set is $\{\frac{-1 - i\sqrt{5}}{6}, \frac{-1 + i\sqrt{5}}{6}\}$.

Problem Set 11.3

1. $\sqrt{81} = 9$ because $9^2 = 81$.

5. $\sqrt[3]{125} = 5$ because $5^3 = 125$.

9. $\dfrac{\sqrt[3]{64}}{\sqrt{49}} = \dfrac{4}{7}$

13. $\sqrt[5]{-243} = -3$ because $(-3)^5 = -243$

17. $64^{\frac{2}{3}} = (\sqrt[3]{64})^2 = (4)^2 = 16$

21. $4^{\frac{5}{2}} = (\sqrt{4})^5 = (2)^5 = 32$

25. $-27^{\frac{1}{3}} = -\sqrt[3]{27} = -3$

29. $(\frac{2}{3})^{-3} = \dfrac{1}{(\frac{2}{3})^3} = \dfrac{1}{\frac{8}{27}} = \dfrac{27}{8}$

33. $125^{\frac{4}{3}} = (\sqrt[3]{125})^4 = (5)^4 = 625$

37. $(\frac{1}{32})^{\frac{3}{5}} = \left[\sqrt[5]{\frac{1}{32}}\right]^3 = [\frac{1}{2}]^3 = \frac{1}{8}$

41. $3^{\frac{4}{3}} \cdot 3^{\frac{5}{3}} = 3^{\frac{4}{3}+\frac{5}{3}} = 3^{\frac{9}{3}} = 3^3 = 27$

45. $\dfrac{3^{-\frac{2}{3}}}{3^{\frac{1}{3}}} = 3^{-\frac{2}{3}-\frac{1}{3}} = 3^{-1} = \dfrac{1}{3^1} = \dfrac{1}{3}$

49. $\dfrac{7^{\frac{4}{3}}}{7^{-\frac{2}{3}}} = 7^{\frac{4}{3} - (-\frac{2}{3})} = 7^{\frac{4}{3} + \frac{2}{3}} = 7^{\frac{6}{3}} = 7^2 = 49$

53. $a^{\frac{2}{3}} \cdot a^{\frac{3}{4}} = a^{\frac{2}{3} + \frac{3}{4}} = a^{\frac{8}{12} + \frac{9}{12}} = a^{\frac{17}{12}}$

57. $(4x^{\frac{2}{3}})(6x^{\frac{1}{4}}) = 24x^{\frac{2}{3} + \frac{1}{4}} = 24x^{\frac{8}{12} + \frac{3}{12}} = 24x^{\frac{11}{12}}$

61. $(5n^{\frac{3}{4}})(2n^{-\frac{1}{2}}) = 10n^{\frac{3}{4} + (-\frac{1}{2})} = 10n^{\frac{3}{4} + (-\frac{2}{4})} = 10n^{\frac{1}{4}}$ 65. $\left(5x^{\frac{1}{2}}y\right)^2 = (5)^2\left(x^{\frac{1}{2}}\right)^2 (y)^2 = 25xy^2$

69. $\left(8x^6y^3\right)^{\frac{1}{3}} = (8)^{\frac{1}{3}}\left(x^6\right)^{\frac{1}{3}}\left(y^3\right)^{\frac{1}{3}} = 2x^2y$

73. $\dfrac{48b^{\frac{1}{3}}}{12b^{\frac{3}{4}}} = 4b^{\frac{1}{3} - \frac{3}{4}} = 4b^{\frac{4}{12} - \frac{9}{12}} = 4b^{-\frac{5}{12}} = \dfrac{4}{b^{\frac{5}{12}}}$

77. $\left(\dfrac{3x^{\frac{1}{3}}}{2x^{\frac{1}{2}}}\right)^2 = \left[\dfrac{3x^{\frac{1}{3} - \frac{1}{2}}}{2}\right]^2 = \left[\dfrac{3x^{-\frac{1}{6}}}{2}\right]^2 = \dfrac{9x^{-\frac{1}{3}}}{4} = \dfrac{9}{4x^{\frac{1}{3}}}$

Problem Set 11.4

1. $y = x+2$ Any real can be substituted for x and all reals will be obtained for y. Therefore, the domain and range both consist of the set of real numbers.

5. $y = x^3$ Any real can be substituted for x and all reals will be obtained for y. Therefore, the domain and the range both consist of the set of real numbers.

9. $y = x^2+2x-1$ The domain is the set of real numbers since any real number can be substituted for x.

13. $f(x) = 2x+7$ The domain is the set of real numbers since any real number can be substituted for x.

17. $y = \dfrac{3x}{x^2-4}$ The denominator, x^2-4, equals zero when $x = 2$ or -2; therefore, the domain is the set of real numbers except 2 and -2.

21. $f(x) = \dfrac{4}{(x+2)(x-3)}$ The denominator, $(x+2)(x-3)$, equals zero when $x = -2$ or 3; therefore, the domain is the set of real numbers except -2 and 3.

25. $y = \dfrac{-2}{x^2+4x} = \dfrac{-2}{x(x+4)}$ The denominator, $x(x+4)$, equals zero when $x = 0$ or $x = -4$; therefore, the domain is the set of real numbers except 0 and -4.

29. $f(x) = -5x-1$

$f(3) = -5(3)-1 = -16$
$f(-4) = -5(-4)-1 = 19$
$f(-5) = -5(-5)-1 = 24$
$f(t) = -5t-1$

33. $f(x) = x^2-4$

$f(2) = 2^2-4 = 0$
$f(-2) = (-2)^2-4 = 0$
$f(7) = 7^2-4 = 45$
$f(0) = 0^2-4 = -4$

37. $f(x) = \sqrt{x}$

$f(4) = \sqrt{4} = 2$
$f(25) = \sqrt{25} = 5$
$f(12) = \sqrt{12} = 2\sqrt{3}$
$f(18) = \sqrt{18} = 3\sqrt{2}$

41. $f(x) = 3x^2-x+4$ $g(x) = -3x+5$

$f(-1) = 3(-1)^2-(-1)+4 = 8$ $g(-1) = -3(-1)+5 = 8$
$f(4) = 3(4)^2-4+4 = 48$ $g(4) = -3(4)+5 = -7$

Problem Set 11.5

1. $f(x) = 4$ The points determined by $f(0) = 4$, $f(1) = 4$, and $f(-2) = 4$ are all on the horizontal line.

5. $f(x) = \frac{1}{2}$ The points determined by $f(0) = \frac{1}{2}$, $f(2) = \frac{1}{2}$, and $f(-2) = \frac{1}{2}$ are all on the horizontal line.

9. $f(x) = 2x-5$ The points $(0,-5)$, $(\frac{5}{2},0)$, and $(1,-3)$ are on the line.

13. $f(x) = -5x-1$ The points $(0,-1)$, $(-\frac{1}{5},0)$, and $(1,-6)$ are on the line.

17. $f(x) = -x$ The points $(0,0)$, $(1,-1)$, and $(-1,1)$ are on the line.

21. $f(x) = \frac{1}{4}x^2$ This parabola has its vertex at the origin and opens upward. The points $(2,1)$ and $(-2,1)$ should help sketch the parabola.

25. $f(x) = -\frac{1}{4}x^2$ This parabola has its vertex at the origin and opens downward. The points $(4,-4)$ and $(-4,-4)$ can be used to help sketch the parabola.